# TECHNICAL REPORT

T0195501

# Enhancing Small-Business Opportunities in the DoD

Nancy Y. Moore, Clifford A. Grammich,
Julie DaVanzo, Bruce Held, John Coombs,
Judith D. Mele

Prepared for the Office of the Secretary of Defense

Approved for public release; distribution unlimited

NATIONAL DEFENSE RESEARCH INSTITUTE

The research described in this report was prepared for the Office of the Secretary of Defense (OSD). The research was conducted in the RAND National Defense Research Institute, a federally funded research and development center sponsored by the OSD, the Joint Staff, the Unified Combatant Commands, the Department of the Navy, the Marine Corps, the defense agencies, and the defense Intelligence Community under Contract W74V8H-06-C-0002.

**Library of Congress Cataloging-in-Publication Data** is available for this publication.

ISBN 978-0-8330-4580-5

Published 2008 by the RAND Corporation
1776 Main Street, P.O. Box 2138, Santa Monica, CA 90407-2138
1200 South Hayes Street, Arlington, VA 22202-5050
4570 Fifth Avenue, Suite 600, Pittsburgh, PA 15213-2665
RAND URL: http://www.rand.org/
To order RAND documents or to obtain additional information, contact
Distribution Services: Telephone: (310) 451-7002;
Fax: (310) 451-6915; Email: order@rand.org

# Preface

This document was prepared in response to a congressional request for a report identifying impediments to small-business owners in contracting or subcontracting with the Department of Defense (DoD). The DoD in turn authorized RAND to undertake this study in February 2008, and the study was completed in May 2008. As requested, the report includes, among other topics, analyses of available data on

- small-business size thresholds and how these affect the ability of a firm to work for the DoD
- contract bundling
- the distribution of small-business subcontracts between professional services and research and development
- transitioning Small Business Innovation Research programs to procurement
- the effects of the DoD Vendor Pay system on small business
- the effectiveness of the Mentor-Protégé Program
- impediments to the success of businesses that graduate from small-business programs or seek to become larger businesses.

This research was sponsored by the DoD Office of Small Business Programs and was conducted within the Acquisitions and Technology Policy Center of the RAND National Defense Research Institute, a federally funded research and development center sponsored by the Office of the Secretary of Defense, the Joint Staff, the Unified Combatant Commands, the Department of the Navy, the Marine Corps, the defense agencies, and the defense Intelligence Community. This research should be of interest to persons concerned with small-business policy and acquisition policy.

For more information on RAND's Acquisition and Technology Policy Center, contact the Director, Philip Antón. He can be reached by email at atpc-director@rand.org; by phone at 310-393-0411, extension 7798; or by mail at the RAND Corporation, 1776 Main Street, P. O. Box 2138, Santa Monica, California 90407-2138. For more information on this research, contact the project leader, Nancy Moore. She can be reached by email at Nancy_Moore@rand.org or by phone at 310-451-6928. More information about RAND is available at www.rand.org.

# Contents

# Figures

# Tables

# Summary

For several decades the federal government has sought to aid and assist small businesses. These efforts have included congressional establishment of government-wide statutory goals for the federal government to purchase at least 23 percent of all its goods and services from small businesses.

Because the Department of Defense (DoD) purchases about two-thirds of all goods and services the federal government buys, its purchasing practices greatly affect the success of federal procurement policy favoring small businesses. The DoD has had mixed success in meeting the procurement goal. Given the importance of DoD purchases to government-wide small-business procurement efforts, Congress asked the DoD Office of Small Business Programs for an assessment of impediments to small-business owners in contracting or subcontracting with the department. This report fulfills that request. Among issues we consider are the unique needs of the DoD and how they affect opportunities with small businesses, contract "bundling," subcontracting in professional services and research and development, opportunities in the Small Business Innovation Research (SBIR) and the Mentor-Protégé Programs, electronic payment systems, and whether firms "graduate" from the programs or increase in size from "small" to larger businesses as a result of various small-business preferences, including those for procurement.

## Opportunities by Industry

The DoD purchases goods and services in an enormous variety of industries. In addition to traditional defense goods such as armored vehicles and ammunition, the DoD purchases at least $500 million in prime contracts annually from grocery, apparel, pharmaceutical, and construction industries.

Still, more than half of DoD purchases are concentrated in just ten industries. In many of these industries, the small-business share of sales to the DoD is below the small-business share of all industry sales. In some industries, including aircraft manufacturing and engineering services, the two industries in which the DoD spends more money than any other, this is because small businesses are less prevalent than they are elsewhere in the economy. In aircraft manufacturing, for example, small firms account for less than 10 percent of the industry, in contrast to their 51 percent share of the overall gross domestic product. It is also possible that small firms within aircraft manufacturing are less likely to make the type of larger transport or advanced fighter aircraft the DoD needs. In other industries, DoD purchases could perhaps include more small businesses, but further research would be needed before identifying specific opportunities with small businesses within them. In still other industries, particularly those from which the DoD makes fewer purchases, the DoD already exceeds the share that small businesses have of industry sales.

Evolving DoD needs may further constrict the opportunities available to small business. Among its broad categories, DoD purchases from small businesses have traditionally been greatest in military construction and family housing and less prevalent in weapon system procurement. As a result, in future years, should weapon system procurement increase relative to other spending, as it is currently projected to do, the small-business share of overall DoD purchases is likely to diminish.

## Bundling

Federal procurement regulations seek to limit consolidation of contracts into a single contract that is not suitable for award to a small business. Indeed, such bundling has been identified by the President and Congressional leaders as a leading impediment to small-business participation in federal contracting opportunities. The prevalence of such practices in DoD and other federal contracting is difficult to determine. One estimate of bundling contends that more than half of DoD prime contract spending is on bundled contracts. Yet another, noting the few formal protests filed over bundling, maintains that there are far fewer cases of bundling. The DoD's own data on bundling are lacking. Given that consolidation of multiple contracts is likely to remain among the purchasing practices that the DoD seeks to adopt, more accurate data are needed on the contract bundling practices that the government seeks to limit as well as on contract consolidations it can accept.

## Subcontracting

In addition to statutory goals for prime contracting with small businesses, the Federal Acquisition Regulation requires that other-than-small businesses submit a subcontracting plan for each solicitation or contract modification that exceeds $550,000 (or $1 million for construction) and offers subcontracting opportunities. Of particular interest to Congress are subcontracting opportunities in research and development (R&D) as well as in professional services. DoD spending in real terms during the past decade more than doubled in R&D and more than tripled in professional services. The small-business share of DoD prime contract dollars for R&D, especially outside that for defense systems, has decreased in recent years, whereas the proportion of dollars spent on contracts requiring a subcontracting plan has increased. Within professional services, the percentage of DoD contract dollars requiring a subcontracting plan has also increased. As a result, small-business subcontracting opportunities in both areas may be increasing, and small-business opportunities in R&D may be shifting from prime contracts to subcontracts. Unfortunately, direct data on DoD subcontracting in these industries are unavailable for analysis. Currently, there is no centralized database for collecting DoD subcontracting data. Nevertheless, the DoD is deploying the Electronic Subcontracting Reporting System which will have data available for analysis.

## SBIR Program

The DoD uses the SBIR program to involve small businesses in its R&D programs. The program is funded by a 2.5 percent assessment placed on external research budgets. It now provides more than $1 billion annually in funds to be spent with small businesses to develop technological innovations. Congress is particularly concerned with how well SBIR technologies move into acquisition programs. Common impediments to such transition in the DoD and other federal agencies are insufficient technical maturity of projects when SBIR funding ends, lack of funding for further development of immature technology, and, as previous RAND research found, program managers who view the program as a burden rather than as a development opportunity.

## Mentor-Protégé Program

An additional means used to increase small-business participation in federal contract opportunities is the Mentor-Protégé Program in which large prime contractors, or mentors, assist certain categories of small businesses, or protégés, to improve their capabilities (including in organizational structure and technology transfer) as suppliers for federal agencies. Data on the program have been limited, although it appears to be popular among both mentors and protégés and perhaps has helped boost the size of protégé firms. Mentor incentives for the program include reimbursement for program participation or credit toward subcontracting goals; the reimbursement option is preferred and hence may be critical to continued operation of the program. DoD protégés report increased revenues from program participation, with exposure to other federal contracting opportunities also viewed positively. Nevertheless, further research is needed on whether the program contributes directly to its goal of boosting small-business opportunities.

## Electronic Payment Systems

Congress has expressed specific concern about requirements to use Vendor Pay for small businesses and how this may affect small-business participation in DoD contracts. Surveys regarding technology in small businesses likewise suggest that electronic pay systems could be a problem. Yet the effect of DoD's e-commerce systems on small businesses appears to be minimal, and small businesses are offered sufficient support to become acquainted with the system. In fact, DoD's e-commerce systems appear to be an improvement over its manual payment methods, given that small businesses have expressed a preference for electronic payment because of the ease in submitting invoices and receiving payments.

## Small-Business "Graduation"

Among the aims of small-business policy are preserving free competitive enterprise and strengthening the overall economy of the nation. Ideally, small firms helped by federal policies that support these aims will, over time, become larger firms. Identifying the extent of such "graduation" with current data is challenging. Nevertheless, our analysis of small businesses in Central Contractor Registration (CCR) data indicates that relatively few small firms providing goods and services to the DoD have graduated over time. Of contractor identification codes appearing in the data between 1997 and 2007 with contract actions reported as going to small businesses, 44 percent were inactive in 2008, suggesting that such firms were either out of business, were acquired by another firm, or were no longer seeking federal contracts; 43 percent were for businesses that had remained "small"; 4 percent were for contractors that dealt in multiple industries, being small in some but not others; and 9 percent were for contractors that were no longer small. Annual archiving of CCR data to better track revenue, growth, and mergers for businesses over time would help foster a better understanding of small-business growth, graduation, and acquisition. Future research might also focus on how larger or longer contracts can help small businesses grow and "graduate" from procurement preferences.

## Recommendations

This research points to several steps federal policymakers could take to address impediments to small-business participation. Reconsideration by the Small Business Administration of what truly defines a small business may result in inclusion of firms not currently receiving small-business preferences but able to bring to market some of the innovations sought through such policies. This might include recognition of the large scale of production or investment necessary in some industries. In extreme cases, this might even include recognizing as "small" all firms not dominant in an industry. Setting goals by industry rather than across the government similarly could help the DoD foster small-business opportunities in industries most conducive to them. Implementing any remedies would require careful analysis of affected industries, including how trends such as globalization or electronic commerce are affecting market dynamics.

Other initiatives will require more sophisticated data analysis to help determine both how well small-business policies are currently performing in meeting their objectives as well as what these objectives should be. Data on contract bundling and subcontracting are lacking or deeply flawed. There have been no efforts to assess how programs such as Mentor-Protégé have directly contributed to their broader objectives. Without more adequate data, Congress may find it difficult to devise solutions to the impediments we discuss, and DoD managers will have difficulty in implementing any solutions.

# Acknowledgments

We are grateful to our project sponsor, Mr. Anthony Martoccia, Director, Office of Small Business Programs (OSBP), Department of Defense, for his sponsorship and to Ms. Linda Oliver, Deputy Director, OSBP, and Ms. Lee Renna, Assistant Director, OSBP, and our project monitor, for their assistance and enthusiastic support of this research effort. We also thank Michael Caccuitto, Wendy Despres, and Paul Simpkins from OSBP as well as Janice Buffler and Kasey Diaz from Triumph Enterprises, who reviewed early drafts or otherwise helped move this project along.

We thank Mr. Tim J. Foreman, Navy Director, Small Business Programs, and Mr. Ronald A. Poussard, Director, Office of Small Business Programs, Office of the Secretary of the Air Force, and his staff for taking time from their busy schedules to meet with us to share their perspectives on improving small-business opportunities within the Department of Defense. We also want to thank Ms. Peg Meehan, Director of the Defense Logistics Agency Office of Small Business Programs, Ms. Tracey L. Pinson, Director of the Army Office of Small Business Programs, and their staffs for providing additional insights on impediments to small businesses as well as small-purchase data.

We thank the people and organizations that supported this project, namely, Ms. Peggy Butler, Manager, Army Mentor-Protégé Program; Mr. Victor Ciardello, Former Director, Small Business Technology and Industrial Base & Program Manager, DoD Mentor/Protégé Program; Dawn Coulter, Director of eSolutions at the Defense Finance and Accounting Service and her dedicated staff; and Ms. Debbie Cabreira-Johnson, Program Director, Los Angeles County Office of Small Business Procurement Technical Assistance Center. We also thank Mr. Roger D. Jorstad, Director, Statistical Information Analysis Division Defense Manpower Data Center, for his help in answering questions regarding data issues and Mr. Rick Rhoadarmer, Triumph Enterprises, for his help in giving us data definitions.

We also thank our former RAND colleague, Dr. Elaine Reardon, for answering questions related to her earlier work, which we updated for this document, as well as our current RAND colleague, Dr. Edward Keating, for sharing his unpublished work, which we updated for this document, and for answering questions about it.

Finally, we thank Dr. Steve Kelman, Weatherhead Professor of Public Management, Harvard University's John F. Kennedy School of Government, for his responsive review of our draft manuscript. We also thank our RAND colleagues, Dr. Susan Gates, for her quick and insightful reviews of early and final drafts, Dr. John Romley for his review of our draft, and Patricia Bedrosian for her quick edits of sections of this document as it was evolving as well as an edit of the completed manuscript. Last, we thank Donna Mead and Sandy Petitjean for their help with document formatting and tables.

# Abbreviations

| | |
|---|---|
| AAP | Acquisition Advisory Panel |
| CAGE | Commercial and Government Entity |
| CCR | Central Contractor Registration |
| CDR | Contract Deficiency Report |
| D&B | Dun & Bradstreet |
| DCMA | Defense Contract Management Agency |
| DFAS | Defense Finance and Accounting Service |
| DLIS | Defense Logistics Information Service |
| DoD | Department of Defense |
| DUNS | Data Universal Numbering System |
| EDA | Electronic Document Access |
| EDI | Electronic Data Interchange |
| EFR | Electronic File Room |
| eSRS | Electronic Subcontracting Reporting System |
| FPDS | Federal Procurement Data System |
| FPDS-NG | FPDS—Next Generation |
| FSC | Federal Supply Class |
| FY | fiscal year |
| GCPC | Governmentwide Commercial Purchase Card |
| HUBZone | Historically Underutilized Business Zone |
| ICAR | Individual Contract Action Report |
| LA OSB/PTAC | Los Angeles County Office of Small Business and Procurement Technical Assistance Center |

| | |
|---|---|
| NAICS | North American Industry Classification System |
| NASA | National Aeronautics and Space Administration |
| NRC | National Research Council |
| O&M | operations and maintenance |
| OMB | Office of Management and Budget |
| OSBP | Office of Small Business Programs |
| PSC | Product and Service Code |
| R&D | research and development |
| RDT&E | research, development, test, and evaluation |
| SBA | Small Business Administration |
| SBIR | Small Business Innovation Research |
| SBTC | Small Business Technology Council |
| SDB | small disadvantaged business |
| SIAD | Statistical Information Analysis Division |
| SIC | Standard Industrial Classification |
| SUSB | Statistics of U.S. Business |
| TRL | technology readiness level |
| WAWF | Wide Area Workflow |
| WInS | Web Invoicing System |

# CHAPTER ONE

# Small Business and the Department of Defense

Congress has long sought to boost small business. These efforts include a statutory goal for the federal government to spend at least 23 percent of prime contract dollars for goods and services with small businesses, with some suggesting increasing this goal to 30 percent (Weigelt, 2007). Because the Department of Defense (DoD) accounts for approximately two-thirds of federal purchases, its purchasing practices greatly affect the success of federal policy to use small businesses (Small Business Administration Office of Advocacy, 2007; House Small Business Committee Democratic Staff, 2006).

The DoD has had mixed success in meeting this goal (Figure 1.1). Over the past half-century, the proportion of its prime contract dollars spent with small businesses has varied between 15 and 25 percent. (In Figure 1.1, as throughout this document, we use FY 2009 deflators in the fiscal year [FY] 2009 DoD "Green Book" [Office of the Under Secretary of Defense (Comptroller), 2008] to calculate constant dollars in this research. We used the FY 2009 deflators because they are the most recent deflators that DoD has calculated and because they reflect budget numbers the Congress is currently considering.)

Concerned about the continuing access of small businesses to DoD contracting and procurement, Congress recently directed the DoD Office of Small Business Programs to request "a report . . . which identifies the impediments to small business owners to contracting or subcontracting with the Department of Defense" (House Appropriations Committee, 2007). More specifically, the committee requested analyses of

- the effects of small-business threshold size
- contract "bundling"
- distribution of small-business subcontracts between professional services and research and development (R&D)
- transitioning Small Business Innovation Research (SBIR) programs to procurement
- the effects of the DoD Vendor Pay system on small business
- the effectiveness of the Mentor-Protégé Program
- impediments to the success of businesses graduating from small-business qualifications.[1]

The Office of Small Business Programs in turn asked the RAND Corporation in February 2008 to prepare this report. RAND researchers fulfilled this request in May 2008. We did so through analyses of existing literature, Economic Census data, federal contract action data,

---

[1] We covered other topics as well, but some, such as regulatory (Crain and Hopkins, 2005) and tax impediments (De Rugy, 2007), proved to be too broad to cover with the time and resources allotted.

**Figure 1.1**
**Total DoD Prime Contract Awards and Percentage of Dollars Going to Small Business,**
**FY 1955 to FY 2006**

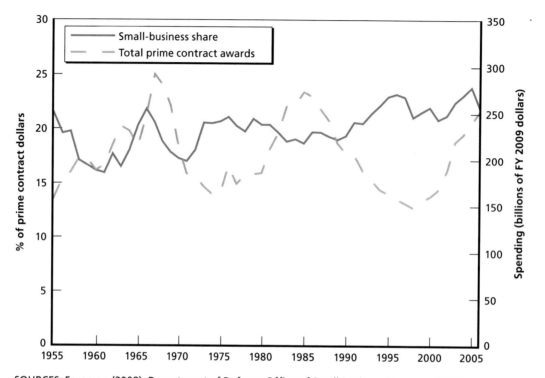

SOURCES: Foreman (2008); Department of Defense Office of Small Business Programs (2008).
RAND TR601-1.1

changes in small-business thresholds over time, and still other data whose quality, and hence the inferences that can be drawn from them, vary. (See Appendix A for a complete discussion of data sources and quality issues.) The breadth of topics requested by Congress, as well as the varied quality of available data on them, in some cases limited the depth at which we were able to explore them.

In the next chapter, we assess the prevalence of small business by industries in which the DoD purchases goods and services and past and projected trends in DoD procurement and how they affect small-business opportunities. We also note some possible adverse effects of threshold size in some industries.

In the third chapter, we examine evidence on contract bundling, the consolidation of requirements into a single contract unlikely to be suitable for a small business because of the larger size of the bundled contract.

In the fourth chapter, we review data on DoD subcontracting in R&D and professional services and note more general concerns about DoD subcontracting.

In the fifth chapter, we examine how well SBIR technologies have moved into DoD acquisition programs, including follow-on research or production contracts with the DoD or its prime contractors.

In the sixth chapter, we review data and literature on the DoD Mentor-Protégé Program in which a large business, acting as a mentor, works with a small firm, or protégé, tutoring it in product development and how to negotiate the federal marketplace.

In the seventh chapter, we discuss the Vendor Pay system and other automated payment systems and issues that small businesses might confront in using them.

In the eighth chapter, we review issues that small businesses contracting with DoD may face in growing to medium-size or larger businesses.

In the ninth chapter, we offer conclusions and recommendations for policymakers and future research.

# CHAPTER TWO

# DoD Prime Contract Purchases from Small Businesses as Defined by Government Threshold Sizes

The U.S. government seeks to provide "the maximum practicable opportunity" for small businesses "to participate in providing goods and services" to it (Small Business Administration, 2008a). To foster this, the Small Business Act has set a government-wide goal to spend "not less than 23 percent of the total value of all prime contract awards for each fiscal year" (Public Law 85-536, as amended, 2004).

To meet the government-wide goal, each agency sets specific goals with the Small Business Administration (SBA). For FY 2007, these ranged from 4.62 percent for the Department of Energy to 60.00 percent for the SBA (Small Business Administration, 2008b). The goal for the DoD was 23.00 percent, the same for 28 other agencies (of a total of 60).

There is an implicit assumption in small-business policy that the federal government should be buying from small businesses in the same proportion that all others do (Clark, Moutray, and Saade, 2006). Assuming that small businesses, which produce about half the gross domestic product (Small Business Administration Office of Advocacy, 2007), should receive a share of federal procurement dollars commensurate with their importance in the U.S. economy requires assuming that the mix of goods and services that the federal government requires is similar to the mix of goods and services produced by the entire economy.

Yet many of the goods and services the DoD requires are relatively unique. DoD purchases, such as those for major weapon systems that it alone uses among American purchasers, tend to be concentrated in industries where small businesses are less prevalent than they are in other industries. This is in great part because such businesses require a large scale of production, one larger than any "small" business can be expected to have.

In this chapter, we examine DoD purchases from small businesses in select industries. We also discuss the changing role of small businesses in these industries and its implications for size standards. We conclude by examining projected trends in DoD budget categories for coming years and what these mean for small-business opportunities.

## Small Businesses in DoD Industries

The DoD purchases an enormous amount and variety of goods and services from a wide variety of industries. Among industries in which the DoD spent at least $500 million in prime contract dollars in FY 2007 are traditional defense industries such as armored vehicles, shipbuilding, aircraft manufacturing, and ammunition manufacturing, as well as industries such as facilities support services, wholesale drugs and druggists' sundries, grocery wholesalers, highway and street construction, apparel, advertising agencies, and single-family housing construction.

In some of these industries, consolidation may be reducing small-business opportunities. Traditionally, defense industries have been less concentrated than civilian high-technology industries (Kopač, 2006). Since the end of the Cold War, firms in these industries have concentrated to levels seen elsewhere in the economy and for similar reasons: to avoid duplication, pool resources, and expand market share. (See also General Accounting Office, 1998a, on defense industry consolidation.)

To assess opportunities for small businesses in defense and other industries with the DoD, we analyzed data on DoD spending by industry and on the prevalence of small businesses by industry in the overall economy.[1] We compared DoD spending by industry and firm size with data on industry and firm size in the U.S. economy, calculating the prevalence of small businesses in different industries of interest to the DoD, noting both industries that provide little opportunity to the DoD and those where the DoD may be able to expand its purchases.

To assess small businesses by industry, we used Economic Census data collected quinquennially by the Census Bureau. These data report by industry the number of firms or establishments[2] in the industry and the number of employees and revenues by firm or establishment size among other variables. We calculated the share that small businesses have in industries as identified by six-digit codes of the North American Industry Classification System (NAICS).

Federal policy uses two criteria to determine small-business eligibility for procurement and other small-business preferences. First, such a firm must be an

Entity organized for profit, with a place of business located in the United States, and which operates primarily within the United States or makes a significant contribution to the U.S. economy through payment of taxes or use of American products, materials, or labor (Small Business Administration, 2008c).

Second, such a business must not exceed the numerical size standard for its industry. For about half of all industries, the government uses the number of employees a firm has to determine whether it is "small." For most industries with an employee size standard, 500 employees is the threshold for determining whether a firm is "small," but thresholds can vary from 100 employees, used for firms in many wholesale industries, to 1,500 employees, used for aircraft manufacturers, among others. For nearly all the remainder, the government uses average annual receipts to determine whether a firm is small. For industries with a size threshold based on average annual receipts, $6 million is the most commonly used threshold for determining whether a business is small, but these thresholds may also vary from $750,000, used for many agricultural industries, to $32.5 million, used for single businesses providing multiple facilities support services.[3]

---

[1]  Much of the research in this chapter follows earlier, similar research by Reardon and Moore (2005).

[2]  The term "establishment" refers to a separate operating location. Thus, a small business typically has one or a few establishments, whereas a large business may have many establishments. It is similar to what is called a contractor in federal contracting.

[3]  For a recent list of industry size thresholds, see Code of Federal Regulations (2008). For our analysis of Economic Census data in 2002, we use thresholds current as of January 1, 2003 (Code of Federal Regulations, 2003). (Thresholds as posted for January 1, 2002, rely on NAICS 1997 codes rather than the NAICS 2002 codes used in the Economic Census of 2002.) Two exceptions to the use of employment size or annual receipts for industry size standards are utilities, for which

Economic Census data classify firms or establishments by aggregate employee or revenue size categories. The use of these aggregate data posed two problems for our efforts to determine the prevalence of small businesses in individual industries. First, category boundaries do not always match thresholds. For example, in some industries, the small-business size threshold is 750 employees, falling in the middle of an aggregate Economic Census employee size category ranging from 500 to 999 employees. In such cases, we assumed that firms and revenues are evenly distributed throughout a category when calculating small-business shares of the industry. Second, for some industries, data in some size categories are suppressed to protect the confidentiality of firms falling in one or more categories with few other firms. We imputed data for such categories from patterns evident in these categories in broader industrial sectors denoted by two-digit NAICS codes. In addition to these two challenges, we note that for some industries, particularly in manufacturing, Economic Census data are available only for establishments, more than one of which may be in a single firm, and not for firms to which small-business size thresholds apply. For these industries, we use establishment data as a proxy for firm data. (For more on these strategies, their assumptions, and their likely effects, see Reardon and Moore, 2005.)

Economic Census data are the best available for assessing small businesses by industry, but, as noted, they are collected only quinquennially. The most recent available data are those for 2002. Results for the 2007 Economic Census will not be available until 2009 or 2010. For some industries, we therefore also cite County Business Pattern data on the distribution of firms by industry with less than or at least 500 employees. To assess consolidation possibly affecting small businesses, we also review "Statistics of U.S. Businesses" compiled by the Census Bureau for the SBA on establishment births, deaths, expansions, and contractions by industry (U.S. Census Bureau, 2008).

To assess DoD purchases from small and other businesses by industry, we analyze data from Individual Contract Action Reports. The Federal Acquisition Regulation requires that each government agency collect data on all contract actions more than $3,000 in value.[4] Within the DoD, these data have traditionally been known as DD350 data, named for the form on which they are collected, and have funneled to the Federal Procurement Data System (FPDS) recording information on all federal contract actions. Within both the DoD and the rest of the federal government, these data are now captured in the FPDS—Next Generation (FPDS-NG), named for the new system used to collect them. These data include, among other data elements, the amount of the action, identification codes including whether the selling firm is a small business, the NAICS code of the firm's industry, and the Federal Supply Class (FSC)

---

small businesses are defined by megawatt hours, and financial institutions, for which small businesses are defined by assets. Because the Economic Census does not include data on megawatt production or bank assets, we do not include these industries in our analyses.

[4] This threshold has varied over time. Before FY 1983, it was $10,000. Between FY 1983 and FY 2004, it was $25,000. In FY 2005 and FY 2006, it was $2,500. Since FY 2006, the Federal Acquisition Regulation has set the threshold at $3,000, excepting a $2,000 threshold for construction work covered by the Davis-Bacon Act and a $2,500 threshold for service work such as custodial, janitorial, or housekeeping services. Any threshold for contract action data can, of course, cause analysts of such data to overlook some small-business opportunities. Unfortunately, although we know that smaller contract actions are more likely to go to small businesses, we have no way to estimate the proportion of all dollars going to small businesses falling below the $3,000 threshold. Unpublished RAND analyses of detailed purchasing data for the Defense Logistics Agency suggest that perhaps 5 percent of that agency's expenditures in FY 2004 were for purchases of less than $2,500, about 44 percent of which went to small businesses, but these results cannot be generalized across years, much less to other DoD agencies.

or Product and Service Code (PSC) for services—more finely grained indicators than NAICS codes provide on the exact nature of goods and services purchased.

In assessing likely future trends of DoD purchases and their implications for small businesses, we also used forecast expenditures by categories such as weapon system procurement and operations and maintenance as published in the National Defense Budget Estimates for FY 2009, also known as the "Green Book" (Office of the Under Secretary of Defense [Comptroller], 2008). We aggregated procurement data by FSCs and PSCs to match Green Book budget categories, identifying small-business shares of these categories and how shifting the distribution of funds among them will likely affect overall DoD procurement from small businesses.

To assess impediments that the structure of DoD purchases may pose to small businesses, we sought to document the share that small businesses have in industries in which the DoD spends the most money, how DoD purchases with small businesses in these industries compare with other federal purchases, and whether consolidation in these industries might be affecting small-business opportunities.

Altogether, the DoD purchases goods and services in more than 1,000 industries. Nevertheless, our analysis of purchases by industry shows that more than half of the $281 billion in DoD spending for external goods and services (as measured by small-business "goaling" dollars) occurs in just ten industries (see Table 2.1, listing, for each industry, total and small-business expenditures by the DoD, in constant billions of dollars benchmarked to FY 2009).[5]

The top categories of DoD expenditures are relatively consistent over time. Total DoD expenditures have increased in all these industries in recent years. Most of these industries were in the top ten in FY 2002. DoD expenditures with small businesses have increased in eight of them and have increased faster than expenditures with other-than-small businesses in five of them.

Small businesses are not as prevalent in these industries as they are in other industries in which the DoD buys goods and services. In nine of the top ten industries in which the DoD purchases goods and services, the overall small-business share of the industry is below the 51 percent share the SBA estimates small businesses have of the nonfarm private gross domestic product (see Table 2.2, showing, for the top ten DoD industries, the small-business share of DoD purchases in FY 2002 and FY 2007 as well as the small-business share of the total industry in 2002).

Conceivably, the DoD may have an opportunity to increase its purchases from small businesses in industries where such businesses have a lower share of the DoD market than of the entire industry. If DoD purchases from small businesses in the industries identified in Table 2.1 were to match industry averages, then DoD procurement from small businesses in these industries would nearly triple, from 12.0 percent to 33.0 percent.

Nevertheless, without knowing additional details about these industries and DoD purchases, it is difficult to estimate what the size of small-business opportunities with the DoD

---

[5]  Table 2.1 and similar subsequent tables are based on small-business goaling dollars and not on all dollars the DoD may spend in an industry. Small-business goaling dollars are dollars that federal authorities consider when determining whether an agency is meeting small-business prime-contracting goals. Among other categories, they exclude foreign military sales, purchases from other foreign concerns, purchases from educational or nonprofit organizations, and purchases from Federal Prison Industries, Inc. (also known as UNICOR). Table 2.1 also lists FY 2007 data that, as of May 2008, had not been officially validated. (DoD subsequently submitted certified data to the Office of Management and Budget in June 2008, after the date DoD specified for completion of this research.) For more on small-business goaling dollars, see Affourtit (2003) and Global Computer Enterprises, Inc. (2006).

**Table 2.1**
**Total and Small-Business Expenditures by the DoD, by Industry (billions of FY 2009 dollars)**

| Industry | Total Expenditures | | Small-Business Expenditures | |
|---|---|---|---|---|
| | 2007 | 2002 | 2007 | 2002 |
| Aircraft Manufacturing | 29.33 | 13.41 | 0.67 | 0.25 |
| Engineering Services | 28.13 | 17.85 | 3.84 | 2.85 |
| R&D in the Physical, Engineering, and Life Sciences | 25.55 | 16.68 | 5.29 | 3.80 |
| Military Armored Vehicle, Tank, and Tank Component Manufacturing | 11.67 | 2.40 | 0.49 | 0.08 |
| Ship Building and Repairing | 10.21 | 10.01 | 0.79 | 0.81 |
| Commercial and Institutional Building Construction* | 9.22 | 5.14 | 3.48 | 2.51 |
| Other Aircraft Parts and Auxiliary Equipment Manufacturing[a] | 8.94 | 7.86 | 1.80 | 1.30 |
| Petroleum Refineries | 6.94 | 2.64 | 0.95 | 0.45 |
| Search, Detection, Navigation, Guidance, Aeronautical, and Nautical Systems | 6.82 | 6.08 | 0.55 | 0.29 |
| All Other Professional, Scientific, and Technical Services*[b] | 5.52 | 2.36 | 0.40 | 0.64 |
| Total | 142.34 | n/a | 18.26 | n/a |

NOTES: Procurement data for FY 2002 used NAICS 1997 codes, not all of which are comparable to NAICS 2002 codes. (Data for FY 2007 used NAICS 2002 rather than NAICS 2007 codes.) For some industries, denoted by an asterisk, we therefore used FY 2003 data. Totals are not provided for FY 2002 because information on procurement data relies on two different years. Industry information was missing for contract actions totaling $9.33 billion (FY 2009) in FY 2007. n/a = not applicable.

[a] This includes manufacturing for goods not included in the Aircraft Manufacturing category as well as those not in another industry category for aircraft engine and engine parts manufacturing.

[b] This is a residual category for services not included in any other six-digit NAICS code for Professional, Scientific, and Technical Services.

might be. We do not know, for example, whether the DoD is in the market for the kinds of airplanes that small businesses manufacture. Perhaps small businesses in that industry produce lightweight general aviation planes less likely to be used by the DoD than by other consumers. Similarly, perhaps armored vehicle manufacturers produce goods for a market quite different from the DoD, such as couriers. Furthermore, in one industry, commercial and institutional building construction, DoD purchases from small businesses may already meet or exceed those of the total economy, possibly leaving it less room to further expand its purchases from small businesses.[6]

---

6   Whether the DoD has actually exceeded industry parity in commercial and institutional building construction is difficult to assess given the nature of Economic Census data. The SBA's "small" business threshold in this industry in 2002 was $28.5 million in annual sales for firms. The top aggregate size category reported in the Economic Census is $10 million or more in value of work done by establishments. We calculate that 23.6 percent of establishments fall below this threshold, and we use this proportion as a proxy for the small-business share of industry in Table 2.2. Such an estimate, though the best available, is imperfect for two reasons. First, of course, the $10 million threshold is below the SBA's $28.5 million size threshold for determining "small" businesses. Second, because a single firm above the threshold may have two establishments below it, relying on establishment data, as must be done in this industry, produces an overestimation of the small-business share of the industry. Reardon and Moore (2005), using data from the 1997 Economic Census, similarly showed

**Table 2.2**
**Small-Business Share of the DoD Market and Industry**

| Industry | Small-Business Share of DoD Purchases, % | | Small-Business Share of Industry, %, 2002 |
|---|---|---|---|
| | 2007 | 2002 | |
| Aircraft Manufacturing | 2.3 | 1.8 | 8.7 |
| Engineering Services | 13.6 | 16.9 | 20.3 |
| R&D in the Physical, Engineering, and Life Sciences | 20.7 | 20.0 | 49.1 |
| Military Armored Vehicle, Tank, and Tank Component Manufacturing | 4.2 | 3.7 | 80.6 |
| Ship Building and Repairing | 7.7 | 8.2 | 32.5 |
| Commercial and Institutional Building Construction | 37.7 | 48.8[a] | 23.6 |
| Other Aircraft Parts and Auxiliary Equipment Manufacturing | 20.1 | 16.1 | 43.7 |
| Petroleum Refineries | 13.7 | 15.0 | 81.5 |
| Search, Detection, Navigation, Guidance, Aeronautical, and Nautical Systems | 8.0 | 4.7 | 31.1 |
| All Other Professional, Scientific, and Technical Services | 7.3 | 27.0[a] | 58.1 |
| Total | 12.0 | n/a | n/a |

NOTE: n/a = not applicable.

[a] This number is based on FY 2003 rather than FY 2002 data.

DoD officials have previously noted difficulties that thresholds pose in areas such as manufacturing, business services, and construction (Hofman, 2006). A typical defense contract in many industries can push a firm over the size threshold. As one DoD representative told us, the scale of a DoD requirement can "bump up against" the size threshold. This can result in "NAICS shopping," that is, searching for an industry code in which a business can still qualify as small and provide the goods or services to DoD. Others have also noted the difficulty that size standards pose in information technology industries (Gerin, 2005). Many such industries have a threshold of $23 million in revenues, and no industry, as noted, has a threshold of more than $32.5 million. As a result, one industry advisor notes, "When you're a $50 million or $100 million company, you're too small to compete with the large boys, and too large to compete in the small-business arena" (Gerin, 2005).

DoD officials also noted to us the difficulties of assessing small-business capabilities in industries in which the DoD purchases the most goods and services. In particular, one told us, "urgent" contracts or short appropriation terms can lead purchasers to large firms rather than taking time to investigate the capabilities of smaller ones. Another noted that without the time to develop a "strategic plan" for maximizing small-business participation, buyers are not able to maximize small-business participation, adding, "it won't work just by giving us more percentages."

It is unclear how consolidation elsewhere in the economy may be affecting opportunities for small businesses within leading DoD industries. Data on establishments show some con-

---

high levels of purchasing by the DoD from small businesses in commercial and institutional building construction. The NAICS code for that industry has since changed and cannot strictly be compared to subsequent data.

solidation in the top industries for DoD purchases. Between 1998 and 2004, Census Bureau data on the number of establishments by industry show that the number of establishments in the top 15 industries in which the DoD buys goods and services decreased 0.1 percent per year whereas the total number of establishments in all industries increased 0.9 percent per year. This suggests that some consolidation may be occurring in top DoD industries and that other industries are decentralizing.

More recently, County Business Pattern data indicate that the proportion of firms with fewer than 500 employees decreased between 2003 (the first year for which data are available for NAICS 2002 codes) and 2005 (the most recent year for which data are available) in four of the top ten DoD industries: ship building and repairing; petroleum refineries; search, detection, navigation, guidance, aeronautical, and nautical systems; and all other professional scientific and technical services. As Table 2.2 indicates, the small-business share of DoD purchases fell in three of these four industries (as well as in engineering systems and commercial and institutional building construction).

SBA thresholds within these industries have remained fairly stable over time, not changing for six of the top ten between 2002 and 2007. SBA thresholds increased in the other four industries: engineering services; commercial and institutional building construction, petroleum refineries[7]; and all other professional, scientific, and technical services. In each of these industries, as Table 2.2 indicates, the proportion of DoD dollars going to small businesses decreased during this time. Put another way, of the five top industries in which the small-business share of DoD dollars decreased, four were industries in which the SBA size threshold increased— and, presumably, the number of small firms increased, all other things equal (although, as mentioned above, there was a decrease in the number of firms with fewer than 500 employees among petroleum refineries). This may indicate that although policymakers recognize the need to increase thresholds in these industries, they might not have increased them enough.

To help identify other issues that the DoD might face in increasing small-business participation, we identified industries in which the DoD spends more than $100 million annually but less than 1 percent with small businesses (Table 2.3). In most of these industries, small businesses are less prevalent than they are in the economy as a whole. In some, including guided missile and space vehicle manufacturing as well as light truck and utility vehicle manufacturing, it may be the case that small manufacturers do not produce the goods and services that the DoD needs. Further investigation would be needed to confirm this. Data may also sometimes be misleading. For example, contracting data show the DoD spending only small portions of its money for nonscheduled chartered freight air transportation, an industry that is virtually all small-business (excepting offshore marine air transportation services firms with annual revenues above $23.5 million). This may be because many of these contracts are reported as being with FedEx, which is the "lead" carrier for broader contracts for transportation services actually performed by other carriers.

In other industries, the small-business share of DoD purchases far exceeds that in the overall economy. In fact, in six industries in which DoD spends at least $100 million, the small-business share of DoD purchases exceeds 95 percent (Table 2.4). In only three of these industries does the small-business share of the industry exceed the 51 percent share small businesses have of the overall economy. In some of these industries, particularly that for janitorial

---

7   More precisely, the employee threshold for petroleum refineries remained at 1,500 employees but the number of barrels that "small" firms were permitted to process daily increased from 75,000 to 125,000.

**Table 2.3**
**Industries in Which the DoD Spent More Than $100 Million in FY 2007 and Less Than 1 Percent with Small Business (millions of constant FY 2009 dollars)**

| Industry | Total DoD Purchases, FY 2007 | DoD Purchases from Small Businesses (in FY09 $millions) | | Small-Business Share of Industry, %, 2002 |
|---|---|---|---|---|
| | | 2007 | 2002 | |
| Guided Missile and Space Vehicle Manufacturing | 5,109.2 | 14.7 | 73.4 | 10.3 |
| Light Truck and Utility Vehicle Manufacturing | 3,119.1 | 27.6 | 8.4 | 8.8 |
| Nonscheduled Chartered Freight Air Transportation | 1,646.4 | 12.2 | 153.4 | 100.0 |
| Animal (Except Poultry) Slaughtering | 356.4 | 0.7 | 0.3 | 23.2 |
| Medicinal and Botanical Manufacturing | 252.0 | 0.2 | 0.7 | 55.4 |
| Scheduled Freight Air Transportation | 195.1 | 1.5 | 0.3 | 43.8 |
| Industrial Design Services | 110.4 | 0.4 | 0.4 | 59.5 |

services and landscaping services, high levels of DoD purchases from small businesses may be a result both of finding small local providers who can tailor their services to local operating needs and of efforts by purchasing officers to increase DoD small-business purchases in areas where they are able to do so. One DoD official claimed that operating bases, having more time to investigate small businesses for services that can serve local needs, "are doing 50 to 90 percent" of their spending with small businesses. Investigating such industries further may offer insights into other areas where small-business opportunities with the DoD can increase. Perhaps most noteworthy is the DoD spend in fruit and vegetable markets, which was about $300 million in FY 2007 but virtually nothing in earlier years. This is largely the result of an indefinite-delivery, small-business setaside contract awarded to Military Produce Group in 2006.[8]

## Future Trends

How might evolving DoD needs affect small-business opportunities? That is, given what is currently known about likely DoD needs in coming years, what are the DoD's prospects for meeting small-business contracting goals? To analyze this question, we examine operational needs that are likely to affect the types of goods and services that the DoD purchases, the small-business share of each, and how the changing compositions of goods and services and the DoD's changing needs will affect the small-business opportunities.[9]

---

[8]  The NAICS code for these contracts also may have been miscoded. That for fruit and vegetable markets is 445230, for which the "small"-business size threshold is $6 million in annual revenues, or more than the average contract for these goods in FY 2007. Military Produce Group is also listed in the Central Contractor Registration as working in NAICS 424480, fresh fruit and vegetable merchant wholesalers, for which the size threshold is 100 employees and not revenue-based. Another significant "small"-business supplier in this industry is Four Seasons Produce, Inc., which has nearly 500 employees. Hardy (2007) notes that the SBA does not routinely check contract NAICS classifications and that a non-manufacturer's rule allows resellers of other companies' products to be considered small if they have no more than 500 employees.

[9]  Much of the research presented in this section is based on earlier unpublished work by our RAND colleague Edward G. Keating.

**Table 2.4**
**Industries in Which the DoD Spent More Than $100 Million in FY 2007 and More Than 95 Percent with Small Business (millions of constant FY 2009 dollars)**

| Industry | Total DoD Purchases, FY 2007 | Percentage of DoD Purchases from Small Businesses | | Small-Business Share of Industry, %, 2002 |
|---|---|---|---|---|
| | | 2007 | 2002 | |
| Fruit and Vegetable Markets | 310.3 | 100.0 | 0.0[a] | 59.1 |
| Janitorial Services | 269.6 | 97.0 | 87.2 | 57.0 |
| Sporting Goods Stores | 205.8 | 100.0 | 57.4 | 42.2 |
| Landscaping Services | 164.3 | 96.4 | 98.1 | 70.4 |
| Industrial Supplies Merchant Wholesalers[b] | 107.6 | 97.7 | 71.7 | 38.9 |
| Photographic Equipment and Supplies Merchant Wholesalers | 101.6 | 96.4 | 59.3 | 17.2 |

[a] DoD had no purchases in this NAICS category in FY 2002.

[b] Because of NAICS comparability problems, DoD purchase data are for FY 2003.

We consider five categories of goods and services that the DoD purchases: operations and maintenance (O&M, excluding pay); weapon system procurement; research, development, test, and evaluation (RDT&E); military construction; and family housing. Expenditures in constant FY 2009 dollars for these have varied over time (Figure 2.1). Procurement was the largest of these categories in the 1980s, but O&M has been the largest since 1991. Spending for O&M, procurement, and RDT&E dwarfs that for military construction and family housing. Spending for all categories is expected to decrease in coming years.

As the composition of DoD expenditures has changed, so, too, has the proportion going to small businesses. For example, in 1985, as weapon system procurement reached its highest level in decades, the proportion of prime contract spending with small businesses decreased to 18.7 percent, its lowest level in decades. In 1996, when weapon system procurement was reaching its lowest level in decades, and O&M spending had surpassed it, prime contract spending with small businesses reached 23.2 percent, its highest level in decades. In 2005, after a rapid increase in O&M spending, which was then exceeding procurement spending by nearly $70 billion, prime contract spending with small businesses reached a still higher level, 23.8 percent.

Using contract action data to approximate small-business spending in each category, we see some variation in the small-business share of each over time (Figure 2.2).[10] Military con-

---

[10]  Our analysis of contract action data classified categories as follows:

- RDT&E: PSCs starting with the letter A.
- military construction: PSCs starting with
  - C1, Architecture and Engineering Services for Construction but excepting C116, Architecture and Engineering Services for Residential Construction
  - Y11, Construction of Administrative Buildings
  - Y12, Construction of Airfields, Communication, and Missile Facilities
  - Y15, Construction of Industrial Buildings
  - Y17, Construction of Warehouse Buildings
  - Y22, Construction of Highways, Roads, Streets, Bridges, and Railways

struction and family housing expenditures have historically had relatively high proportions of dollars going to small businesses, but these proportions have been decreasing over time. In fact, in recent years, they have been close to the proportion of O&M dollars going to small businesses. Small-business use in the other three, larger categories, including O&M, has been more stable. A higher proportion of O&M dollars than of RDT&E or weapon system procurement dollars has been going to small businesses. The proportion of RDT&E expenditures going to small businesses gradually increased through 2000 but has decreased since then. Weapon system procurement has had the smallest proportion of its prime contracting dollars go to small businesses in the past two decades—less than 10 percent.

What do these data mean for future small-business use? To examine this question, we applied the FY 2007 small-business use rate for each category of expenditures to the projected total of each category of spending in coming years. Table 2.5 demonstrates this process for FY 2009. The table shows that, overall, we may expect prime contract expenditures with small businesses in these categories to be $67 billion, or 19.8 percent of the $338 billion total for these categories, assuming that small-business shares of each do not change from their FY 2007 levels.

The Green Book for FY 2009 provides projections by category through FY 2013, as shown in Figure 2.1. We used these to simulate the small-business share of prime contract expenditures in these five categories through that year. We calculate that the overall share of small-business prime contract purchases in these categories will decrease in coming years, especially as the share of total expenditures for weapon procurement, where small-business utilization is relatively less, increases after FY 2009 (Figure 2.3). By 2013, if small-business shares of each category remain as they were in FY 2007 (and do not, for example, continue to decrease for military construction and family housing), then we expect the small-business share of prime contract expenditures in these categories to decrease below 19.2 percent. This would represent the lowest level of prime contract purchases from small businesses since the 1980s (compare Figure 1.1). The primary reason for this will be the concentration of more DoD dollars in sectors of the economy in which small business traditionally has been less concentrated, with

---

- family housing: PSCs starting with
  - C116, Architecture and Engineering Services for Residential Construction
  - E161, Purchase of Family Housing
  - X161, Lease or Rental of Family Housing
  - Y161, Construction of Family Housing
  - Z161, Maintenance, Repair, and Alteration of Family Housing
- weapon system procurement: FSCs and PSCs starting with
  - 1, Aircraft and Systems
  - 2, Components
  - 4470, Nuclear Reactors
  - 58, Communication Equipment
  - 660, Navigational Instruments
  - 661, Flight Instruments
  - 662, Engine Instruments
  - H, Quality Control, Testing, and Inspection
  - K, Modifications
  - N, Installation of Equipment
  - W, Lease and Rental of Equipment
- O&M: all other PSCs and FSCs.

**Figure 2.1**
**Actual and Projected DoD Expenditures, by Category, FY 1980 to FY 2013**

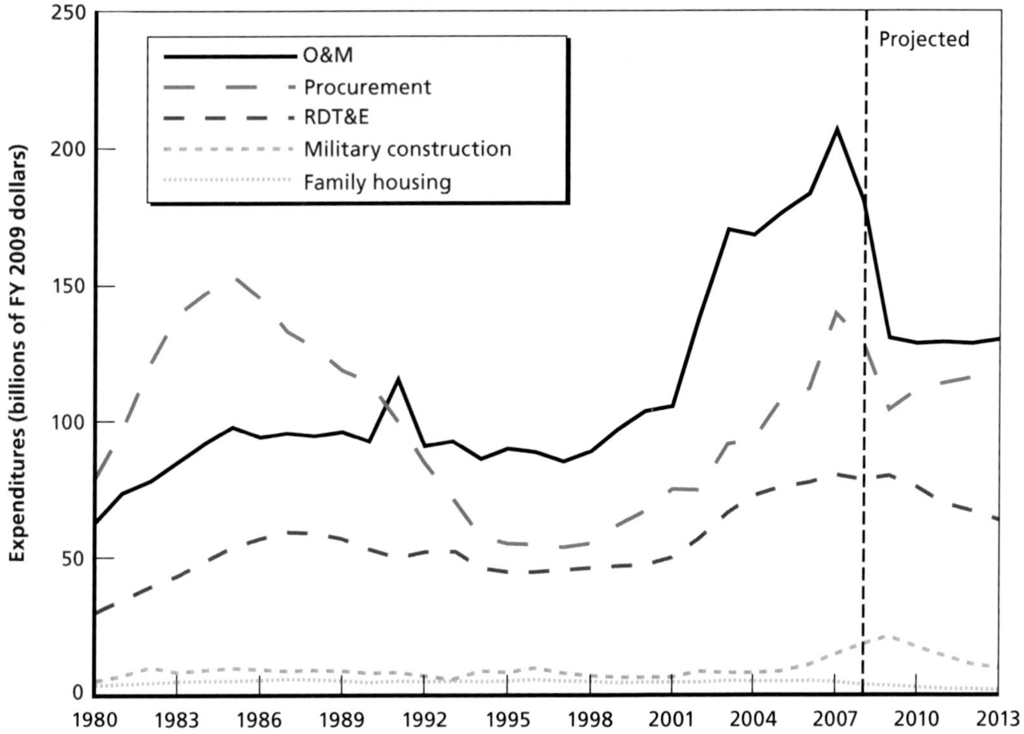

SOURCE: Office of the Under Secretary of Defense (Comptroller) (2008).
RAND *TR601-2.1*

relatively high levels of weapon system procurement spending in each year leading to relatively fewer purchases from small businesses.[11]

## Conclusion

In this chapter, we have examined past and likely future trends of opportunities for small businesses to contract with the DoD. DoD spending with small businesses varies by the industries in which it spends money, as well as by its shifting priorities and needs. Any initiatives to increase small-business opportunities to deal with the DoD must acknowledge both these realities.

Timely analyses of future Economic Census data as they are released may be helpful in assessing and identifying future small-business opportunities by industry. The Economic Census allows identification of changes in small business by industry, although such analysis can be made difficult by changes in industry classification and codes over time. Assessing such data as they are published may help the SBA adjust its thresholds to economic changes affecting small-business opportunities by industry.

---

[11] Many private enterprises face similar problems in seeking to increase purchases from minority suppliers. There are few minority enterprises in the steel industry, for example, making it difficult for manufacturers that have a great demand for steel products to meet diverse spending goals (Duffy, 2004).

**Figure 2.2**
**Use of Small Businesses as Prime Contractors, by Budget Category, FY 1980 to FY 2007**

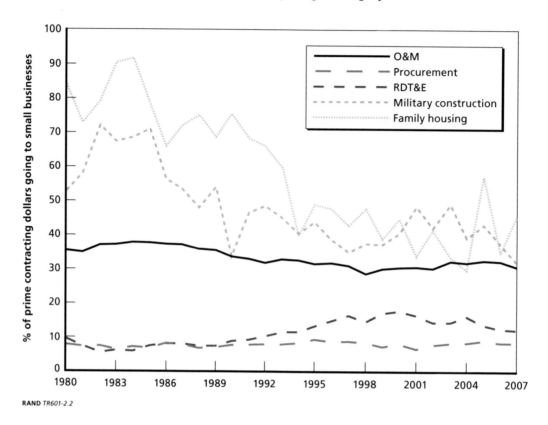

RAND *TR601-2.2*

**Table 2.5**
**Projected Total and Simulated Amount of Prime Contract Expenditures Going to Small Business, by DoD Budget Category, FY 2009 (billions of dollars)**

| Category | Proposed Amount, by Category, per Green Book (A) | Amount Going to Small Business, per FY 2007 Contract Action Data (B), % | Estimated Amount Going to Small Business (A × B) |
|---|---|---|---|
| O&M (excluding pay) | 129.957 | 30.66 | 39.841 |
| Procurement | 104.216 | 8.58 | 8.938 |
| RDT&E | 79.616 | 12.44 | 9.901 |
| Military Construction | 21.197 | 32.10 | 6.804 |
| Family Housing | 3.284 | 45.46 | 1.493 |
| Total | 338.270 | 19.80[a] | 66.977 |

[a] Calculated by dividing the sum of simulated dollars ($66,977 million) by the sum of Green Book projections for these categories ($338,270 million).

**Figure 2.3**
**Weapon Procurement and Total Prime Contract Awards to Small Businesses as a Percentage of DoD Expenditures in Five Leading Budget Categories, Estimated and Projected, FY 2007 to FY 2013**

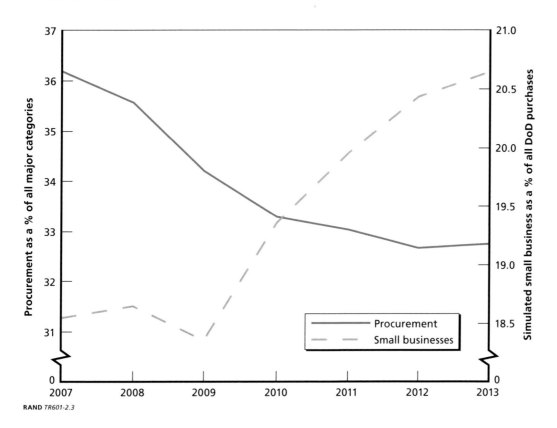

RAND *TR601-2.3*

# CHAPTER THREE

# The Effect of Bundling on Small-Business Opportunities

Among the small-business issues of greatest concern to federal policymakers in recent years has been the practice of contract bundling. This is the consolidation of

> Two or more procurement requirements for goods or services previously provided or performed under separate, smaller contracts into a solicitation of offers for a single contract that is unlikely to be suitable for award to a small business concern (Public Law 85-536, as amended, 2004).

Both the President and congressional leaders believe that reducing bundling would increase opportunities for small businesses. President Bush has said, "I believe the best way to help our small businesses is . . . to unbundle government contracts so people have a chance to bid and receive a contract to help get their business going" ("Transcript of Debate Between Bush and Kerry, with Domestic Policy the Topic," 2004). Senator John Kerry, currently the Chairman of the Senate Committee on Small Business and Entrepreneurship Committee, has suggested that a link exists "between increases in contract bundling and a decline in small business contracting" and has vowed "to reduce the effects contract bundling is having on small businesses" (U.S. Senate Committee on Small Business and Entrepreneurship, 2005). Similarly, Representative Nadia Velázquez, currently the Chairwoman of the House Small Business Committee, maintains that "something must be done about the rampant contract bundling that is driving small businesses out of the federal marketplace" (Velázquez, 2003).

Concern over bundling of federal contracts has arisen as commercial practices change to consolidate more purchases and the federal government has sought to adopt many of the best commercial practices in its own purchases. In this chapter, we review commercial practices on consolidating purchases, federal adoption of these practices and their implications for small-business opportunities, and evidence on the prevalence of bundling in DoD contracts.

## Commercial Practices

Commercial firms in recent years have sought to consolidate their purchases and sales of products. Common examples of consolidated product offerings in the commercial sector include communications companies offering cable television, Internet, and telephone services at a combined price lower than what they offer separately for these services; jet engine manufacturers offering engine parts and repair at a combined lower price; and facilities management services offering janitorial, grounds-keeping, plumbing, electrical, and other building maintenance services together. Common examples of commercial firms consolidating their purchases include soliciting bids for enterprise-wide raw materials such as steel, aluminum, or fuel; manufacturing inputs such as bearings, circuit cards, or motors; or services such as cell phone use, medical insurance, or temporary services.

Commercial practice has changed in response to two influences. The first, prevalent in supply chain purchases, is the success that Japanese automakers and, subsequently, firms in other industries have had in achieving superior quality, responsiveness, and lower total costs through supply chain transformation (Womack, Jones, and Roos, 1991; Womack and Jones, 1996; Laseter, 1998; Nelson, Mayo, and Moody, 1998; Liker, 2004).

Traditionally, manufacturers have used a batch-and-queue mass production supply chain (Figure 3.1). In this system, each individual supplier provided batches of goods or services directly to the inventory of the manufacturer, which in turn assembled the final product from raw materials, piece parts, and components. Because such a system involved a very large and sometimes overwhelming number of direct, short-term relationships with suppliers, it provided little opportunity for the manufacturer to focus its efforts to improve quality and responsiveness such as those to reduce variance in inputs, synchronize production, and reduce costs throughout the supply chain. It is particularly difficult to efficiently integrate a very large number of suppliers into product design, manufacturing, or supply chain operations. (For more on the adverse effects of a large number of suppliers on quality, see Trent, 2001.)

As manufacturers have sought to reduce waste through "lean" practices such as minimal inventory, "just-in-time" supply, and use of fewer, larger, and more complex assemblies, they have also sought to use a smaller, more stable supply base that is well integrated with product

**Figure 3.1**
**Notional Illustration of Batch-and-Queue Mass Production Supply Chain**

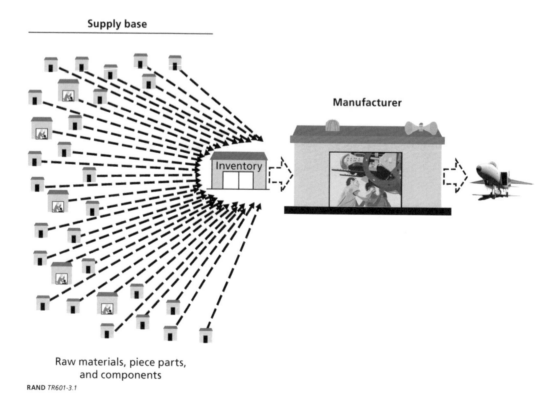

design and synchronized with manufacturing (Figure 3.2). Goods and services continuously flow from lower-tier suppliers to a smaller number of larger, higher-tier suppliers. This permits the manufacturer to focus its efforts to improve quality and responsiveness and reduce total costs with a small number of Tier 1 suppliers (that is, suppliers that provide a larger amount of goods to it directly). The manufacturer then asks its Tier 1 suppliers to do the same with their Tier 1—the manufacturer's Tier 2—suppliers. See Figure 3.3 for an example of a lean supply chain—the Boeing Dreamliner's 14 Tier 1 suppliers and the major subassemblies of the aircraft for which they are responsible.

Leading commercial firms, and the federal government, have similarly sought to develop strategic sourcing—that is, a collaborative and structured process of analyzing an organization's expenditures and using the subsequent information to make business decisions about acquiring commodities and services more effectively and efficiently (Johnson, 2005)—and reduce their number of direct contracts and suppliers. Focusing on longer-term relationships with these suppliers can also improve quality in the supply chain. (For more on the adverse effects of short-term relationships on the supply chain, particularly their costs, see Hahn, Kim, and Kim, 1986.)

Leading commercial examples of large manufacturers who have consolidated purchases from their suppliers include Procter and Gamble, Cisco, Intel, and Boeing. Procter and Gamble is developing enterprise-wide spending pools and applying strategic sourcing to increase

**Figure 3.2**
**Notional Illustration of a Lean Supply Chain**

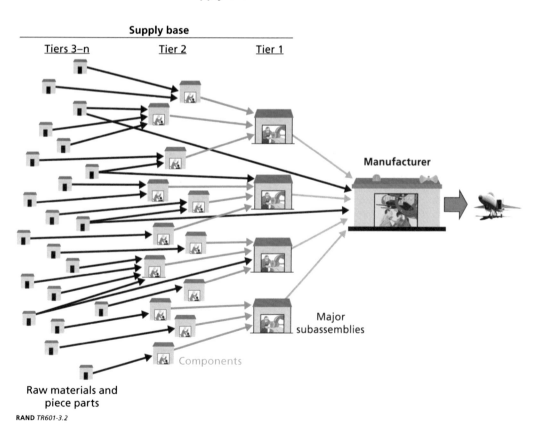

RAND TR601-3.2

**Figure 3.3**
**Boeing's 787 Lean Supply Chain Supplier Partners**

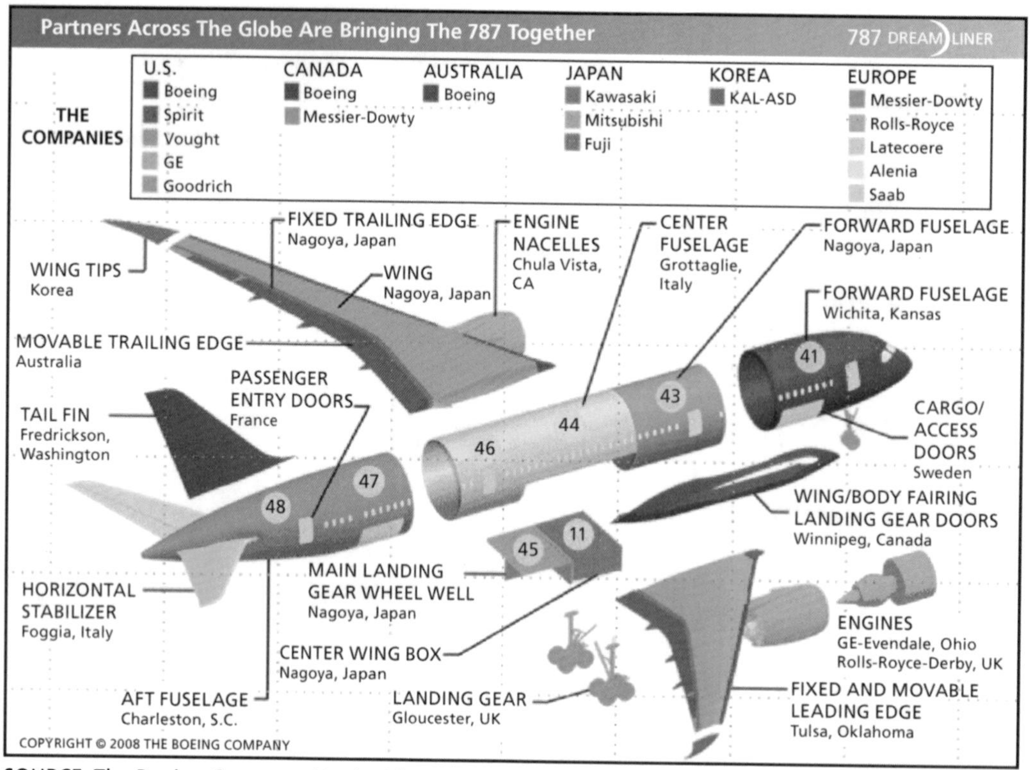

SOURCE: The Boeing Company. Used with permission.
**RAND** *TR601-3.3*

leverage with large key suppliers (Smock, 2004). Cisco has developed a strategy for each commodity group it buys, detailing the volume of parts and the number of and flexibility with suppliers it needs, subsequently reducing its number of suppliers from 1,300 to 300 ("At Today's Cisco Systems, the Fewer Suppliers the Better," 2006). Although fostering diversity programs for its U.S. suppliers with certification from the SBA, Intel seeks to avoid having small amounts of business with large numbers of small suppliers and instead to have its diverse suppliers grow in size (Atkinson, 2008). Such consolidation need not lead to reduced overall opportunities for small businesses. In constructing its 787 "Dreamliner," for example, Boeing has sought to function "less as a manufacturer than as a project manager, supervising its first- and second-tier contractors, each of which may rely on scores of more specialized subcontractors" (Hise, 2007). Altogether, Boeing has used fewer Tier 1 suppliers but at least as many small, lower-tier suppliers, more than 900, on the 787 as it did on the 777 (Hise, 2007). Further, each larger Tier 1 supplier is contracting out bigger jobs to their own Tier 1 suppliers.

Efforts like these have led or are leading to the consolidation of many industries. For example, defense industries (e.g., those concentrating in major weapon systems) that had been more decentralized than others have become more consolidated (Kopač, 2006). Large DoD suppliers have cut their base of subcontractors by about 50 percent in attempts to leverage their purchases, cut costs, and improve supplier performance in providing goods and services (e.g., quality, delivery, and cost) (General Accounting Office, 2004).

The second influence on commercial practices is the grouping of goods and services together into one offering, particularly as a company's goods become more like commodities with lower profits and their services (e.g., repairs) for these goods become more profitable (Auguste, Harmon, and Pandit, 2006; Johansson, Krishnamurthy, and Schlissberg, 2003). Firms such as General Electric and Rolls Royce were among the first to combine service contracts with initial product purchase. For example, Rolls Royce offers, under its trademark "Power-by-the-Hour," a jet engine service contract that guarantees a level of operational performance and charges the customer a fee based on the hours the engine is flown. Similarly, the DoD may expect its leading suppliers to offer more goods and services grouped together for purchase such as those it seeks for performance-based logistics.

## Government Practices

The federal government, including the DoD, has sought to implement many of the same commercial practices that have helped industries reduce total costs and improve performance. The General Accounting Office (now the Government Accountability Office) has recommended that the DoD assess its total contract spending to centralize its purchasing and increase its buying power (General Accounting Office, 2003). The Office of Management and Budget (OMB) has also directed federal agencies "to leverage spending to the maximum extent possible through strategic sourcing" (Johnson, 2005). To reduce their purchasing costs, federal agencies need to improve supplier selection processes, reduce their number of suppliers, leverage major suppliers, maximize economies of scale, and increase automation of transactions (Bail, 2006).

The success of service contracts in the commercial sector was instrumental in leading the DoD to direct program managers to establish a performance-based logistics approach for weapon system support, integrated supply chain management, and other Life-Cycle Logistics[1] responsibilities (Defense Acquisition University, 2004). Changes in defense industries have also led defense agencies to rely on prime contractors to integrate legacy systems and new capabilities in "systems of systems" and to subcontract with new firms that deliver novel capabilities and innovative products (Hayward, 2005). Other factors such as changes in the DoD workforce may also have contributed to this change. In building the F-22 Raptor aircraft, for example, Lockheed Martin performed less than half the work on billable materials and relied on teams of suppliers for much of the work it did not perform itself (King and Driessnack, 2007).

Such practices may have mixed results for small-business opportunities, reducing the number of small businesses receiving prime contracts but possibly providing them the same total dollars. For example, if the DoD were to consolidate ten annual contracts with ten small businesses into one contract with one small business, the total amount of DoD dollars going to small businesses may not change although the number of contracts issued to small businesses and the number of small businesses receiving such contracts would be reduced tenfold.

---

[1] "Life-Cycle Logistics (LCL) is the planning, development, implementation, and management of a comprehensive, affordable, and effective systems support strategy, within TLCSM [Total Life Cycle Systems Management]. Life cycle logistics encompasses the entire system's life cycle including acquisition (design, develop, test, produce and deploy), sustainment (operations and support), and disposal" (Defense Acquisition University, 2008, p. 8).

The reduction would be still greater if the DoD were to consolidate its purchases across years into one contract. Yet even if the number of dollars spent with small businesses were to remain unchanged, implementing best purchasing practices would still lead to a lower number of contracts and to fewer small businesses receiving federal purchasing dollars.

Congress previously took action to reduce contract bundling that might result from such consolidation and reduce opportunities for small businesses. The Federal Acquisition Regulation further states that federal departments and agencies should, to the maximum extent practicable, structure contracting requirements to facilitate participation by and competition among small businesses and avoid unnecessary and unjustified bundling of contractual requirements that may preclude participation by small businesses as prime contractors.

Before bundling any contracts, agencies must demonstrate measurable and substantial benefits such as cost savings, quality improvements, reductions in acquisition cycle times, or better terms and conditions (Baldwin, Camm, and Moore, 2001). More specifically, to be approved, bundled contracts of value up to $86 million must have benefits of at least 10 percent of the contract's total value, and those of more than $86 million must have benefits of 5 percent of the contract's value or $8.6 million, whichever is greater (General Accounting Office, 2004).

Small businesses may, without losing their small-business status, form joint ventures with other firms, large or small, to bid on a (non-setaside) contract (that is, a contract not otherwise set aside for a special category of small business, such as an 8a business) too large for them to handle alone. Although this helps preserve some opportunities for small businesses, creating and sustaining such arrangements can be difficult and will pose many transaction costs and management challenges (Romney, 1998). Some small businesses have also found prime contracts with the federal government more profitable than subcontracts to prime contractors providing bundled goods and services to the federal government (Aitoro, 2006).

## Prevalence of Bundling in DoD Procurement

The extent of bundling and its effect on small-business contracting opportunities with the DoD is not well understood.

The SBA Office of Advocacy sponsored one of the first analyses of bundling by Eagle Eye Publishers (2002). This analysis defined bundled contracts as contracts that included dissimilar goods and services, were performed at varying places, or had varying type-of-contract codes. Using this definition, the analysis found that 10 percent of contracts and 55 percent of the $1.2 trillion spent on defense contracts between 1992 and 2001 were bundled.

The assessments of the General Accounting Office over the years have found more limited evidence of bundling. For example, a 2004 analysis by the General Accounting Office found eight bundled contracts within the Department of Defense in FY 2002, the second-largest number of bundled contracts among federal agencies (General Accounting Office, 2004). This was much lower than the 109 DoD contracts indicated in the FPDS as bundled that year. The analysis noted that the other 101 "bundled" contracts were awarded to small businesses despite statutory language defining bundled contracts as unsuitable for small business.

Although the General Accounting Office analyses suggested an overcount of bundled contracts in the FPDS, the Eagle Eye study suggested that the FPDS data undercount bundling by failing to address "accretive bundling" (Eagle Eye Publishers, 2002). Accretive bundling, Eagle Eye contends, "occurs when contract [sic] officers add new tasks to existing GSA [Government Services Administration] Schedule, Indefinite Delivery/Indefinite Quantity, Government Wide Acquisition Contracts and other multiple award-type contracts" (Eagle Eye Publishers, 2002, p. 4). This, they claim, has become the most widely practiced form of bundling. One of our interviewees indicated that this type of task addition happens when contracting officers are pressed for time to meet a new requirement.

Yet another analysis, relying on different data, suggests that the SBA analyses "materially overstate" the extent of bundling and its effect on small business. Nerenz (2006, 2007) used data on contractor bid protests over cases of bundling to assess its extent and its seriousness in comparison to other issues affecting small-business opportunities. He found that only 18 protests were filed by small businesses over bundling between 1992 and 2001. This suggests a far lower level of bundling than SBA estimates of more than 34,000 cases of bundling that, the SBA contends, forced nearly 15,000 firms from the federal marketplace in these years. The difference, Nerenz maintains, is a result both of the broad definition of bundling Eagle Eye used in its analysis, which did not match the statutory definition, as well as of the lack of actual bundling activity that small businesses felt was detrimental to them.

To improve data on contract bundling and consolidation, in FY 2002 the DoD added a data element to the DD350 form to indicate whether contract requirements had been bundled or consolidated. Table 3.1 shows the number of contracts this indicator identified as bundled between 2001 and 2004.

These data show that less than 0.5 percent of contracts were reported as bundled in these years, as well as less than 0.4 percent of dollars associated with these contracts in three of these years and only 1.31 percent of dollars associated with these contracts in FY 2001. They also show 203 small-business contracts in FY 2001 and FY 2002 as bundled, despite the federal definition of bundled contracts as being unsuitable for small businesses. This suggests that many contracts consolidating multiple contracts with small businesses were being recorded as bundled contracts. The General Accounting Office (2004) found coding errors made on procurement data as a result of confusion about the statutory definition of bundling, inadequate verification of data, and ineffective controls in the reporting process. Those errors appear to have been corrected in FY 2003 and FY 2004, in which small businesses are shown as receiving zero bundled contracts.

In FY 2005, the new FPDS-NG included data elements for both bundled contracts, which small businesses are deemed unable to handle, and "consolidated" contracts, which they can. Table 3.2 shows bundled and consolidated contracts evident in these data.

**Table 3.1**
**Reported Contract Bundling in DD350 Data, FY 2001 to FY 2004**

| Bundling Indicator | FY 2001 | | FY 2002 | | FY 2003 | | FY 2004 | |
|---|---|---|---|---|---|---|---|---|
| | All | Small Business | All | Small Business | All | Small Business | All | Small Business |
| No. of contracts | 378 | 167 | 126 | 36 | 43 | 0 | 25 | 0 |
| % of all contracts | 0.42 | 0.28 | 0.10 | 0.04 | 0.03 | 0 | 0.01 | 0 |
| % of dollars | 1.31 | 0.30 | 0.32 | 0.29 | 0.32 | 0 | 0.30 | 0 |

NOTE: The table presents data only on contracts and contract actions for goaling dollars.

**Table 3.2**
**Reported Contract Bundling in FPDS-NG Data, FY 2005 to FY 2007**

| Bundling Indicator | FY 2005 | | FY 2006 | | FY 2007 | |
|---|---|---|---|---|---|---|
| | All | Small Business | All | Small Business | All | Small Business |
| **Bundled** | | | | | | |
| No. of contracts | 31 | 3 | 208 | 52 | 688 | 181 |
| % of all contracts | 0.00 | 0.00 | 0.05 | 0.02 | 0.18 | 0.23 |
| % of dollars | 0.84 | 0.00 | 0.48 | 0.05 | 0.64 | 0.18 |
| **Consolidated** | | | | | | |
| No. of contracts | 24 | 2 | 5,232 | 3,207 | 17 | 5 |
| % of all contracts | 0.01 | 0.00 | 1.41 | 1.17 | 0.00 | 0.01 |
| % of dollars | 0.01 | 0.00 | 1.49 | 2.15 | 0.28 | 0.05 |

NOTE: The table presents data only on contracts and contract actions for goaling dollars.

These data also show a relatively low number of bundled contracts, but there are several problems evident in them. First, there are discrepancies between these data, which have been reviewed, and what DD350 data indicate for bundled contracts. In FY 2006, for example, whereas the FPDS data shown in Table 3.2 indicate that there were 208 bundled contracts, including 52 to small businesses, DD350 data indicate that there were 379 bundled contracts, including 226 to small businesses.

This may illustrate the problems that contracting personnel are having in accurately recording bundled contracts. Second, as noted, these data show bundled contracts for small businesses that, ostensibly under federal definitions, are unsuitable for such contracts.

In recent years, the DD350 and FPDS-NG have included a data element to indicate exceptions to bundled contracts, that is, contracts that were appropriately bundled for measurably substantial benefits or mission-critical systems (systems whose failure will likely lead to mission failure). Requirements consolidation that results from competing federal civilian jobs for prospective outsourcing as described in OMB Circular A-76 is also exempt from bundling regulations.[2] Table 3.3 summarizes data on contract bundling exemptions.

In each year, mission criticality was cited more often than OMB Circular A-76 as justification for bundling. Other reasons for exempting contracts from bundling restrictions were also cited more frequently in FY 2006 and FY 2007. Discrepancies between FPDS-NG, which was reviewed, and DD350 data in FY 2005 and FY 2006, however, also suggest reporting problems and limit inferences that should be drawn from them.

---

[2]   Office of Management and Budget (2003) Circular No. A-76 established federal policy for the competition of commercial activities, that is, "a recurring service that could be performed by the private sector and is resourced, performed, and controlled by the agency through performance by government personnel, a contract, or a fee-for-service agreement."

**Table 3.3**
**Reported Contract Bundling Exceptions in DD350 and FPDS-NG Data, FY 2002 to FY 2007**

|  | From DD350 | | | From FPDS-NG | | |
| --- | --- | --- | --- | --- | --- | --- |
|  | 2002 | 2003 | 2004 | 2005 | 2006 | 2007 |
| Mission-critical contracts | 26 | 12 | 10 | 10 | 15 | 39 |
| % of contracts | 0.02 | 0.01 | 0.00 | 0.00 | 0.00 | 0.01 |
| % of dollars | 0.04 | 0.10 | 0.09 | 0.16 | 0.12 | 0.05 |
| OMB Circular A-76 contracts | 7 | 3 | 2 | 3 | 4 | 11 |
| % of contracts | 0.01 | 0.00 | 0.00 | 0.00 | 0.00 | 0.00 |
| % of dollars | 0.03 | 0.05 | 0.02 | 0.03 | 0.03 | 0.04 |
| Other exempted contracts | 113 | 59 | 13 | 18 | 189 | 683 |
| % of contracts | 0.09 | 0.04 | 0.01 | 0.00 | 0.05 | 0.17 |
| % of dollars | 0.26 | 0.17 | 0.18 | 0.65 | 0.33 | 0.55 |

## Conclusions

Consolidation of contracts and suppliers in the private sector will continue and is likely to adversely affect the viability of small businesses in the broader commercial economy, particularly in major manufacturing industries such as automotive and aviation. Although total small-business revenue may not go down, requirements of contract consolidation and longer-term contracts within DoD's small-business base will lead to fewer small-business contracts and will bolster concerns about contract consolidation and bundling.

As the General Accounting Office has found, the DoD needs to improve the accuracy of its data on contract bundling and consolidation so that it can better assess its extent. The OMB has suggested developing a scorecard for agencies to use in rating their progress toward unbundling contracts (Hardy, 2006). Other proposals have suggested redefining bundling and requiring that federal officials assess its effects on new contracts. The DoD needs to track where bundling is occurring not only within the DoD but also in the commercial sector in industries where it procures goods and services. More generally, it may wish to consider where small businesses can best contribute to innovation, including at Tier 1 or lower-level suppliers. Tracking bundling or contract consolidation in both the public and private sectors can help inform policy on bundling as well as policies on subcontracting and small-business thresholds.

# CHAPTER FOUR

# Subcontracting in Professional Services and Research and Development

In addition to statutory goals for prime contracting with small businesses, the Federal Acquisition Regulation requires that other-than-small businesses submit a subcontracting plan for each solicitation or contract modification that exceeds $550,000[1] (or $1 million for construction) and offer subcontracting opportunities. The subcontracting plan must document actions that the contractor will take to provide varying types of small businesses (e.g., small "disadvantaged" businesses) with subcontracting opportunities. It should also include the total dollars and types of work to be subcontracted, goals and types of work for each type of small business, and efforts that the contractor will take to meet these goals. For FY 2008, the DoD seeks to ensure that 37.2 percent of dollars for subcontracts—that is, for any agreement entered into by a government contractor calling for subcontracted supplies or services required for contract performance—are spent with small businesses (Department of Defense Office of Small Business Programs, 2008).[2]

Since FY 1989, between 34 and 43 percent of DoD subcontracting dollars have gone to small businesses (Department of Defense Office of Small Business Programs, 2008). As a result, small businesses now receive about $40 billion in subcontract awards from the DoD, or somewhat close to the roughly $50 billion they have received in prime contract awards. Indeed, some analysts contend that "larger participation by small businesses will come at the subcontracting level" (Clark, Moutray, and Saade, 2006; but see also Aitoro, 2006, on the greater possibilities some small businesses may see in prime contracts).

In requesting research on impediments to small-business contracting opportunities with the DoD, Congress specifically requested analysis of subcontracting opportunities in research and development and in professional services. There are several reasons to focus on these areas of subcontracting.

First, as we will see, DoD spending in R&D, as measured in FY 2009 constant dollars, nearly doubled in the past decade, and that for professional services nearly tripled. Both grew much faster than that for all prime contract and subcontract spending by the DoD, which increased only about 60 percent in real terms.

Second, the DoD's purchasing in other areas increasingly encompasses professional services as well. For example, rather than just buying computers from suppliers, the DoD, like

---

[1]  This threshold was $500,000 before FY 2007 when it was adjusted for inflation (Federal Register, 2006).

[2]  The base for determining achievement of subcontracting goals includes all contracts for more than $1 million in construction and $550,000 for all other goods and services, not just those otherwise requiring subcontracting plans (General Services Administration, 2008).

other large purchasers, may now buy a broader range of information services, recognizing a need to improve its processes for purchasing professional services (Gottlieb, 2004).

Third, because small businesses may have "relatively few layers of management and simple organizational structures," the large businesses with which the DoD does most of its contracting may conclude that "they can accomplish their objectives through their small-business subcontractors more easily than through internal resources from large, cross-functional departments" (Dunn, 2004). In particular, small businesses are considered more likely to pursue "radical innovations" and to be "vital to the success of the computer, biotechnology and other high technology industries" (Dobbs and Hamilton, 2007).

In this chapter, we review available data on DoD purchases of professional and R&D services to determine what these indicate about opportunities and impediments for small businesses in these areas. Unfortunately, as we will discuss, DoD subcontracting data afford only limited opportunities for analysis and identification of impediments to small businesses subcontracting in these fields. Nevertheless, extant data and literature point to some broad issues for consideration.

## Data on Prime Contracts and Subcontracts for Professional and R&D Services

To assess small-business participation in providing professional and R&D services, we again turn to DD350 and FPDS-NG data. As noted in our discussion above of small-business prime contract opportunities by industry, these data provide information on, among other variables, the amount of a contract action, whether the seller is a small business, the NAICS code of the seller's industry, and the PSC for these services—a more finely grained indicator than the NAICS code on the exact nature of services purchased. Because the PSC focuses on the exact nature of the services purchased, rather than on the seller's predominant industry, it is better for identifying DoD purchases of R&D and professional services. To assess R&D contacts, we identify all contract actions for RDT&E, that is, for all contract actions with a PSC beginning with A. For professional services, we identify all contract actions for Professional, Administrative, and Management Support Services, that is, for all contract actions with a PSC beginning with R.[3]

### RDT&E

As noted, DoD spending on R&D has nearly doubled in the past decade, increasing most rapidly since 2001. Among RDT&E categories, that for Defense Systems R&D has accounted for most dollars in the past decade, increasing from $14.8 billion to $23.7 billion (Figure 4.1).[4] Nevertheless, rates of growth in RDT&E have been greater in other categories, such as that for General Science and Technology R&D, which increased from $162 million to $2.72 billion, and for Medical R&D, which increased from $47 million to $349 million.

As RDT&E expenditures have grown, prime contracting opportunities for small businesses have changed, particularly outside Defense Systems R&D. Traditionally, small

---

[3]   A complete listing of PSCs is available at http://www.fpdsng.com/downloads/psc_data_05182007.xls (as of May 6, 2008).

[4]   We remind the reader that as of May 2008, data for FY 2007 had not yet been officially validated.

**Figure 4.1**
**DoD Prime Contract Spending on RDT&E, FY 1997 to FY 2007**

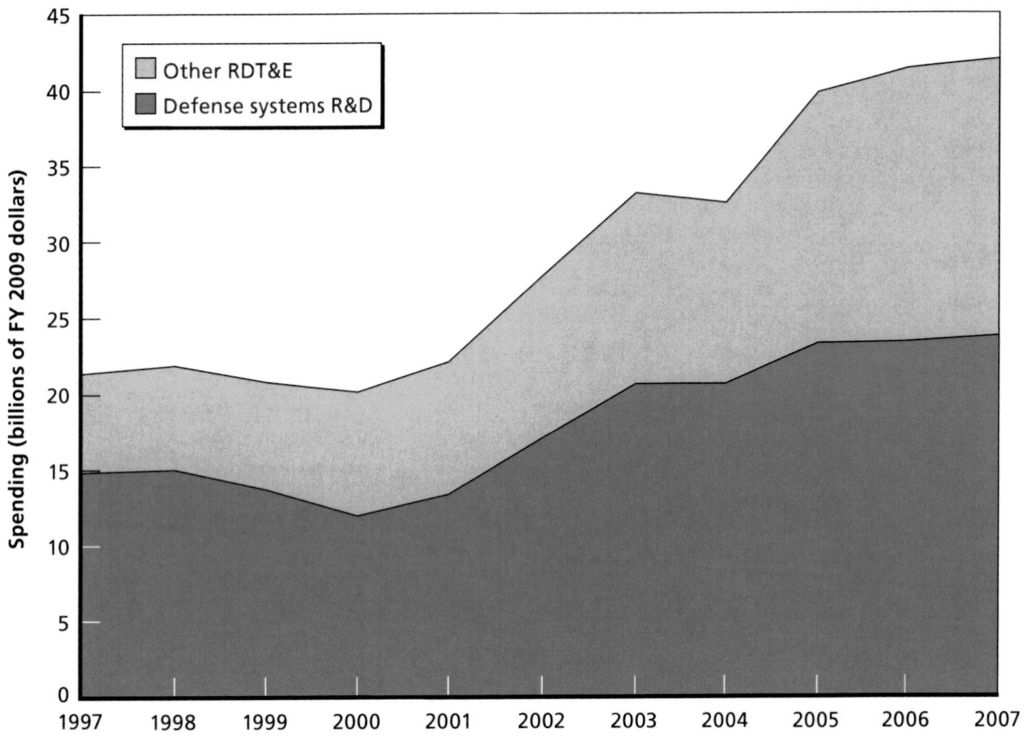

RAND *TR601-4.1*

businesses have won a much higher percentage of prime contract dollars for "other" RDT&E than they have for Defense Systems RDT&E (Figure 4.2). Yet in recent years, as other R&D has grown rapidly, the proportion of contract dollars going to small businesses providing these services has decreased sharply.[5] That for Defense Systems has remained relatively stable. To be sure, given the increasing total number of dollars spent in RDT&E, even with a decreasing proportion going to small business, the total expenditures with small businesses in these industries have risen from $1.55 billion to $2.21 billion (in constant FY 2009 dollars) for Defense Systems R&D and from $2.11 billion to $2.00 billion for all other RDT&E. Again, however, total small-business expenditures in RDT&E have not risen as fast as those expenditures in other businesses.

---

[5]  It is unclear how changing SBA thresholds may have affected this decrease in dollars going to small businesses. Most of the other RDT&E expenditures are in the industries of Engineering Services, NAICS code 541330, and Research and Development in the Physical, Engineering, and Life Sciences, NAICS code 541710. The size thresholds for 541710 are employee-based: 500 employees except for R&D related to aircraft (1,500 employees), aircraft parts and auxiliary equipment (1,000 employees), and space vehicles and guided missiles (1,000 employees). Those for 541330 are revenue-based: $6 million for engineering services firms except $25.0 million in FY 2007 for those providing such services for military and aerospace equipment and weapons and $17.0 million for those firms providing such services for marine engineering and naval architecture. The SBA threshold for all (other) engineering services increased 32 percent in real terms between FY 2000, when SBA thresholds for NAICS (rather than Standard Industrial Classification) codes were first published, and FY 2007. Yet those for military and aerospace equipment increased only 10 percent in real terms and that for marine engineering and naval architecture increased only 11 percent.

**Figure 4.2**
**Small-Business Share of DoD Prime Contract Dollars for Defense Systems R&D and Other RDT&E, FY 1997 to FY 2007**

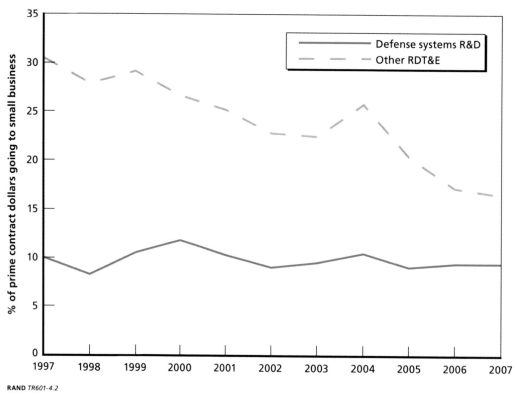

Subcontracting opportunities for Defense Systems R&D, like prime contracting opportunities, appear to have remained stable, whereas those for other RDT&E have fluctuated (Figure 4.3). The proportion of dollars for other RDT&E spent on contracts requiring a subcontracting plan decreased sharply from FY 2002 to FY 2004. In FY 2007, it reached its highest level in at least a decade. It remains below that for Defense Systems R&D; this is not surprising given that other RDT&E still has a higher proportion of prime contract dollars spent with small businesses. Altogether, the value of Defense Systems R&D contracts requiring a subcontracting plan increased from $12.7 billion to $17.6 billion in the past decade, whereas that of other RDT&E contracts requiring a subcontracting plan increased from $4.0 billion to $12.4 billion.

Taken together, the trends evident in Figures 4.2 and 4.3 suggest that small-business opportunities in RDT&E may be developing faster in subcontracting than in prime contracting. The amount of DoD prime contract dollars going to small businesses in RDT&E, especially in the area outside Defense Systems R&D, is not increasing as fast as the amount going to other-than-small businesses in these areas. Dollars for which subcontracting plans are required are keeping pace with total expenditures but the percentage of those dollars for which the plan applies (i.e., those going to subcontracting) varies with each contract. Analysts further considering small-business opportunities in these areas may wish to explore how broader changes in RDT&E are affecting opportunities available to small businesses.

**Figure 4.3**
**Proportion of Dollars for DoD Defense Systems R&D and Other RDT&E Spent on Contracts Requiring a Subcontracting Plan, FY 1997 to FY 2007**

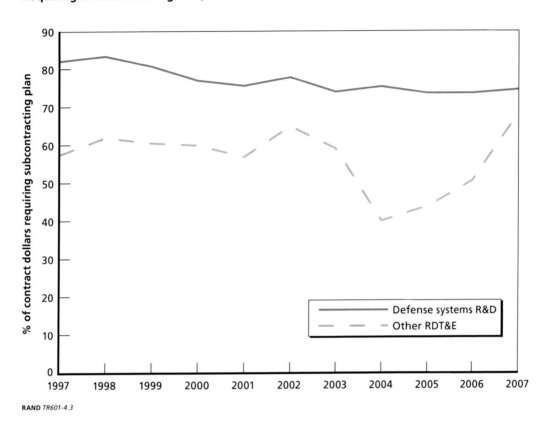

RAND *TR601-4.3*

## Professional Services

DoD spending on Professional Services has increased even faster than that for R&D. The total amount spent for these services still lags that for R&D. Between FY 1997 and FY 2007, the value of DoD prime contracts for Professional, Managerial, and Administrative Support Services increased from $10.1 billion to $30.0 billion (Figure 4.4). The bulk of these expenditures, and the bulk of growth in them, is in Professional Services. Expenditures for this category increased from $7.9 billion to $24.3 billion (in FY 2009 dollars). Expenditures for Managerial Services increased at a slightly slower rate, from $2.0 billion to $5.0 billion. Those for Administrative Support Services increased more rapidly, from $198 million to $727 million.

In contrast to those in RDT&E, small businesses in Professional, Managerial, and Administrative Support Services have largely maintained their proportion of prime contract dollars in the past decade (Figure 4.5). Small-business proportions of Professional and Managerial Services have held relatively steady. Coincidentally, they have also held above government-wide goals for small-business procurement. The small-business share of prime contract dollars for Administrative Support Services has fluctuated more widely but has remained above 50 percent throughout the past decade.

The proportion of DoD dollars spent on contracts requiring subcontracting plans has fluctuated more unevenly but in 2007 were nearly as high as they had been at any point in the past decade (Figure 4.6). In FY 2007, it was the highest it had been in Professional Services

**Figure 4.4**
**DoD Prime Contract Spending on Professional, Managerial, and Administrative Support Services, FY 1997 to FY 2007**

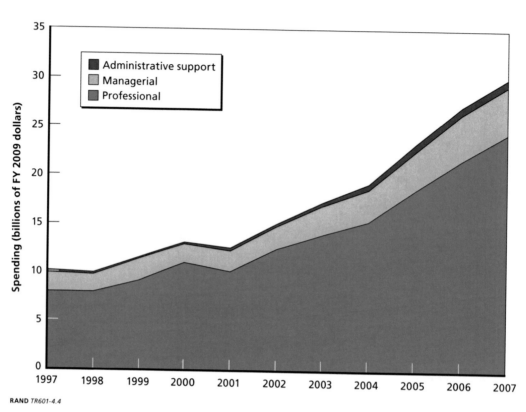

RAND *TR601-4.4*

since FY 2000, in Managerial Services since FY 2002, and in Administrative Support Services in the past decade.[6] Again, however, the percentage of those dollars for which the plan applies (i.e., those going to subcontracting) varies with each contract.

The trends evident in Figures 4.5 and 4.6 indicate that, in contrast to those in RDT&E, small-business opportunities with the DoD in Professional, Managerial, and Administrative Support Services have remained relatively stable in prime contracting as have the amounts of dollars for which subcontracting plans are required. Both prime contracting and subcontracting opportunities for these services may have kept pace with the overall growth in DoD spending for them. Yet it is also the case that the proportion of dollars spent on contracts requiring subcontract plans is a poor proxy for actual small-business subcontracting opportunities in providing these services.

## Other Subcontracting Data

Ideally, to assess subcontracting opportunities in RDT&E and professional services, we would assess direct data on subcontracts rather than subcontracting indicators on prime contract data. Currently, DoD does not collect subcontracting data in a centralized database suitable

---

[6]  Readers should remember that small businesses are not required to file subcontracting plans to enhance small-business subcontracting opportunities. Given the large proportion of Administrative Services contract dollars spent with small businesses, it is not surprising that the proportion of dollars spent on contracts requiring subcontracting plans is relatively low.

**Figure 4.5**
**Small-Business Share of Prime Contract Dollars for Professional, Managerial, and Administrative Support Services, FY 1997 to FY 2007**

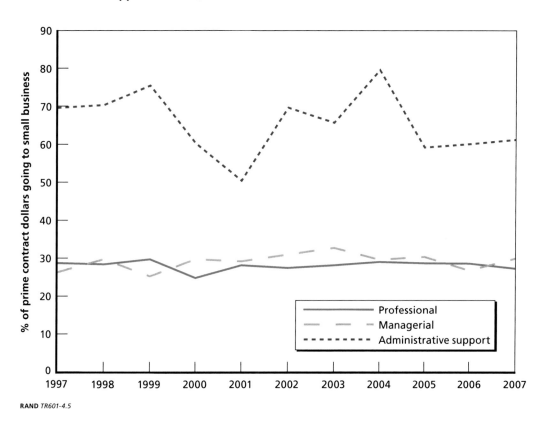

for analysis. DoD is deploying the Electronic Subcontracting Reporting System (eSRS), which should provide suitable data for future analyses.

The Defense Manpower Data Center annually collects and publishes data on Standard Form 295, a Summary Subcontract Report for all subcontracting done by an establishment for an agency and its distribution among special categories of small businesses. Because such data are in summary form, they cannot be used to assess subcontracting for specific goods or services or with specific small businesses. The DoD also collects on Standard Form 294 sub-contracting data for each individual contract requiring a subcontracting plan. Such data could be linked to specific contracts and PSCs, but their collection has not been automated or cen-tralized. Hence, these data are unavailable for analysis.

The President's Management Agenda for Electronic Government has recognized the need for better data on subcontracting. As a result, the SBA and other agencies have collaborated to develop the eSRS to collect the data currently reported on Standard Forms 294 and 295 using Web-based forms (Office of the Under Secretary of Defense for Acquisition, Technology, and Logistics, 2008). The DoD began deploying eSRS in April 2008. Although not available for this analysis, eSRS data could prove beneficial for subsequent analyses of small-business sub-contracting opportunities.

**Figure 4.6**
**Proportion of DoD Contract Dollars for Professional, Managerial, and Administrative Support Services Requiring a Subcontracting Plan, FY 1997 to FY 2007**

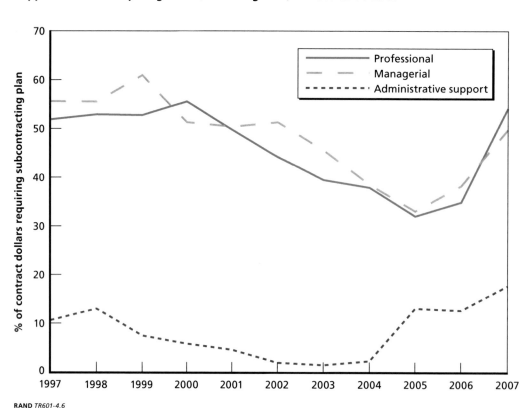

RAND *TR601-4.6*

## Continuing Issues and Implications

As purchases for R&D and professional services increase, the DoD may seek to adopt more commercial practices for these industries. In the commercial sector, subcontracting has been increasing as companies assess activities that are strategic to their operations, determining those they could perform more efficiently in-house and those they may choose to outsource. In some industries such as manufacturing, firms may choose to spend up to 70 percent of their budgets through outsourcing or subcontracting for goods and services ("What's in a Name? New CPO Title Reflects Buyer's Strategic Role," 2000). Different industries may take different approaches depending on industry dynamics, regulatory requirements, company business models, or even historical and cultural roots. For example, among airlines, older firms such as United, American, and Delta perform more of their own heavy maintenance than do newer firms such as Southwest ("Differences of Opinion Fail to Taint Benefits of Outsourcing," 2000).[7]

The exact approach the DoD should take will depend on developing a better understanding of its current practices. We cannot, unfortunately, determine the current extent of DoD subcontracting in specific industries. We can only assess some broad indicators of its possible

---

[7]   For further discussion of strategies that firms may use to subcontract management services, particularly for facilities, see Moore, Grammich, and Bickel (2007).

practice. In seeking to promote small-business subcontracting opportunities, the DoD may need a better understanding of market dynamics that may be favoring larger firms in these industries, particularly in RDT&E, as well as the exact nature of services that can be subcontracted further.

# CHAPTER FIVE

# Obstacles to Technology Transition in the SBIR Program

The Small Business Innovation Research program seeks to stimulate technological innovation, to use small business to meet federal R&D needs, to foster and encourage participation by "disadvantaged" persons in technological innovation, and, ultimately, to increase private-sector commercialization of innovations derived from federal R&D funding (General Accounting Office, 1987). The DoD uses the SBIR program as a way to involve small businesses in its R&D efforts to support the development and procurement of weapons and other equipment and systems that will be used by the armed forces.

The SBIR program has its origins in the Small Business Innovation Development Act of 1982 (Public Law 97-219, 1982). The program receives its funds from a fee assessed on federal agencies with extramural R&D budgets of at least $100 million. Originally set at 1.25 percent, this assessment is now 2.5 percent of the extramural R&D budget. As the level of this assessment has changed and R&D budgets have increased within the DoD, so has the amount of the SBIR budget.[1]

Between 1983 and 2007, program funding, as measured in constant dollars, increased from less than $30 million to nearly $1.2 billion (Figure 5.1). As a result, SBIR funds are an increasingly significant source of funds for small-business R&D, providing the DoD with potentially greater opportunities to tap the innovative capabilities of small businesses.

At the same time, the contribution of SBIR to the increase in small-business participation in RDT&E is unclear. The rate of growth for the SBIR program between FY 1997 and FY 2007, 78 percent, has been greater than the 48 percent increase for other small-business RDT&E. Yet absolute growth in SBIR in real terms, $537 million, has lagged that of other small-business RDT&E, $1.19 billion (Figure 5.2). Overall, in the past decade, DoD RDT&E expenditures with small businesses have increased from $3.49 billion to $5.21 billion.

## SBIR Process

The DoD solicits SBIR research topics several times each year. Small businesses wishing to participate in the program compete to win contracts to conduct the R&D identified in the solicitations. Each SBIR project is awarded in two phases. Phase I awards are generally limited to $100,000 and are used to determine the merit and feasibility of the proposed research. If

---

[1] The growth in the SBIR budget could also reflect an increasing proportion of research, development, test, and engineering funding going to extramural activities as the RDT&E budget increased and the size of the acquisition workforce decreased. Additionally, the increase in the DoD SBIR budget could also occur as DoD agencies hit the $100 million threshold, although instances of this occurring were not specifically sought in the research to date.

**Figure 5.1**
**DoD SBIR Budget and RDT&E Outlays, FY 1983 to FY 2007**

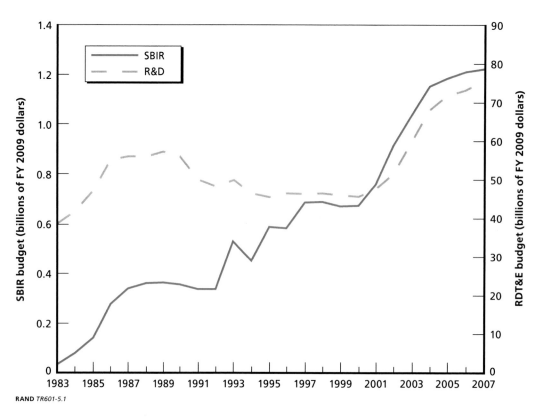

RAND *TR601-5.1*

the Phase I effort demonstrates that the research topic has merit and is feasible, the small business may be invited to propose and win a Phase II contract. Phase II contracts may provide up to $750,000 to continue the work started in Phase I.[2] Phase II includes most of the substantive research on a topic and often results in a prototype or laboratory demonstration of the technology.

A third phase seeks to develop products for commercial or military markets. Phase III funds are not from the SBIR program but from other sources. If the small business has its own resources, it may self-fund further technology development and commercialization. More typically, small businesses approach other sources of funds, such as venture capitalists and other investors for continuation funding. Alternatively, and particularly in the case of the DoD's SBIR program, small businesses may look to continued government support through direct government R&D contracts or through a subcontract with a government prime contractor.

The DoD SBIR program seeks to move technologies into acquisition programs and ultimately the into hands of warfighters. Commercialization of SBIR technologies has also been of increasing concern to Congress.[3] How well DoD SBIR-developed technologies make this transition is not clear. In fact, it appears that only in the last decade have there been serious

---

[2]   The $100,000 Phase I and $750,000 Phase II limits are guidelines. Awards are often smaller and occasionally larger than the stated limits.

[3]   In the 1992 and 2000 reauthorizations of the SBIR program (Public Law 102-564, 1992; Public Law 106-554, 2000), Congress emphasized the importance of the commercialization goal. The 2006 Defense Authorization Act (Public Law 109-163, 2006) authorized the Secretary of Defense and the Secretaries of the Military Departments to establish a Com-

**Figure 5.2**
**Growth in SBIR and Other Small-Business RDT&E Expenditures by DoD, FY 1997 to FY 2007**

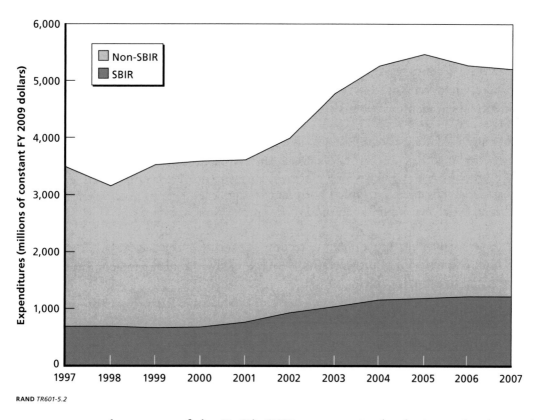

RAND *TR601-5.2*

attempts to assess the success of the DoD's SBIR program in developing technologies that transfer into the DoD acquisition system and to understand the impediments to this process. We review below the small body of literature on these assessments.

## Transfer of SBIR Technology to the DoD Acquisition System

Three principal independent efforts have assessed the DoD's SBIR program, including its success in technology transfer. Those of the General Accounting Office (now the Government Accountability Office) are the oldest, having examined the SBIR program almost since its beginning. Those of the National Research Council (NRC) of the National Academies have been the largest and include a series of studies, symposia, and conferences starting in the late 1990s. Finally, those of the RAND Corporation, starting in 2004, have also focused on impediments to technology transfer and commercialization.

### General Accounting Office Evaluations

Of these efforts, those of the General Accounting Office are the least specific in evaluating the DoD SBIR program's technology transition efforts and the obstacles such efforts face. Early General Accounting Office assessments noted that SBIR programs in the DoD and National

---

mercialization Pilot program within the SBIR program. Sales to the government or continued Phase III government R&D contracts or subcontracts are considered commercialization.

Aeronautics and Space Administration (NASA) differed from those in other departments by stressing mission needs when funding projects (General Accounting Office, 1989). Other SBIR programs, such as those at the National Science Foundation and the Department of Health and Human Services, focused more on R&D projects within broad categories of interest that would result in commercial success.

Regarding technology transfer, the General Accounting Office (1989) noted only that the "SBIR program helps transfer technology by creating networks among SBIR contractors, government, and academia." Yet it also observed

> DoD and NASA are using SBIR projects to undertake high-risk research—research in areas where results are less easy to achieve. In these two agencies, about half of the Phase II SBIR projects were rated by project officers as having higher levels of risk than non-SBIR projects that they managed. Only 13 percent of the projects in these agencies were assessed as having lower levels of risk than comparable non-SBIR projects.

This suggests one possible impediment to technology transfer: the use of the SBIR program to conduct high-risk research of projects that were less likely to make it to acquisition regardless of the source of funding for them.

Subsequent research also contrasted the DoD focus in SBIR projects on mission needs with the focus of most other departments on private-sector commercialization. A report of the General Accounting Office (1992) noted that, "As a matter of policy, DoD is also the only major agency that is emphasizing federal R&D needs in contrast to private-sector commercialization." The report found that only 40 percent of DoD SBIR-generated technology led to private-sector sales, a lower percentage than that achieved by SBIR programs in all other participating federal departments. The report further suggested that the DoD SBIR program focus on dual-use technologies to bolster private-sector commercialization.

Another report of the General Accounting Office (1997) was more positive about DoD SBIR transitions. It noted that 52 percent of SBIR follow-on sales were to the DoD or its prime contractors, indicating "the research results are being used in military projects and programs." Neither the 1992 nor the 1997 report mentioned possible impediments to the transition of SBIR technologies into DoD acquisition programs.

A later report of the General Accounting Office (1998b) briefly discussed commercialization of DoD SBIR projects but it did not specify whether commercialization efforts were the result of further contracts with the DoD or its prime contractors. It also said nothing about impediments that DoD SBIR contractors faced when trying to win Phase III work sponsored by the government. More recent analyses have focused on program administration and management and have said nothing about commercialization or about success and impediments to technology transition.[4]

Although the General Accounting Office has not specifically addressed impediments to technology transition in the SBIR program, it has examined the problem more generally across DoD technology development programs. For many years (most recently in Government Accountability Office, 2006b), it has advocated best practices that delay technology transition until the product is mature enough to be developed. Premature technology transition

---

[4]   One of these reports (Government Accountability Office, 2006a) discussed award data at the DoD and the National Institutes of Health. Another (Government Accountability Office, 2006c) focused only on the collection and reporting of program and commercialization data.

creates technical, cost, and schedule risks. If the DoD SBIR program focuses its research at the basic and applied levels,[5] acquisition program managers seeking to reduce their program risks are likely to view those technologies with suspicion, or as being less ready for development and more likely to increase risks, creating a significant impediment to transition. The General Accounting Office also identified needs for planning, a gated technology development process,[6] transition agreements defining expectations, and managers to work the transition process between the science and technology and acquisition communities. NRC and RAND analyses have noted many of these same needs.

## National Research Council Evaluations

The NRC has undertaken the most extensive examination of the DoD SBIR program. Its initial effort recognized the difficulty that small businesses have in SBIR technology transition. An early NRC evaluation (Wessner, 1999, p. 49) quotes a congressional staff member on the possible need for "partnering and support . . . to make sure that those small businesses are getting into the supply chain and the programs for the Federal Government" as they sought to navigate the "schizophrenic" program and its sometimes contradictory requirements to support the needs of the sponsoring department and to succeed in the commercial marketplace. This echoes the findings of the General Accountability Office on the contrast between the DoD emphasis on having the program meet mission needs and that of other departments on commercial success.

Early NRC research (Wessner, 1999) also commented on the DoD SBIR program's Fast Track Pilot, now a continuing program. The Fast Track program provides additional funding to DoD SBIR companies that also secure outside funding for Phase II research. This, a DoD official said, helps many companies that were good at writing proposals for SBIR contracts and doing useful R&D but not at turning research into products that could be sold to the DoD. The Fast Track program provides a way for small businesses to find others to assess the value of the research and possibly support its transition.

In a formal review of the Fast Track Pilot program, the NRC suggested that a time lag between Phase I and Phase II of the program creates a cash flow problem for small companies and poses a potential impediment to SBIR technology transition (Wessner, 2000). The NRC also noted that the Fast Track program helped firms bridge this gap by helping them attract additional research funding. It also found that technology being developed in the Fast Track Pilot was closer to commercialization and that firms involved in the pilot program experienced greater sales than did firms not in the program. The NRC did not consider the source of sales,

---

[5]  The DoD groups research and technology development into seven categories. Basic and applied research is the most fundamental and describes research that is aimed at understanding basic scientific principles and how to apply them to technological solutions (Department of Defense, 2006).

[6]  The Government Accountability Office (2006b) report describes gated processes in place at several successful product-oriented companies, including well-defined criteria to measure whether a particular step is complete. When the managers of the process decide to move to the next step, they submit to a review or "gate." If the completion criteria are met, then the process moves through the gate and on to the next step. This is very similar to the stage-gate process that is more commonly known (Cooper and Edgett, 2008):

> A Stage-Gate System is a conceptual and operational road map for moving a new-product project from idea to launch. Stage-Gate divides the effort into distinct stages separated by management decision gates. Cross-functional teams must successfully complete a prescribed set of related cross-functional tasks in each stage prior to obtaining management approval to proceed to the next stage of product development.

so it was not clear whether Fast Track technologies had faster transitions into DoD acquisition programs.

Subsequent analyses (Wessner 2007a, 2007b), conducted by congressional request, focused on the DoD SBIR program and Phase III commercialization. Regarding Phase III commercialization, the NRC notes

> In theory, there would be a smooth flow of technology and funding from Phase II to Phase III and then into systems eventually adopted by the agencies for use by warfighters. In reality, this process is much more complex, requiring multiple champions at different phases in addition to effective management and product development by the SBIR firm. The process can, and does, work. There are important success stories. Nonetheless, there are substantial barriers that impede Phase II projects from successfully transitioning into Phase III (Wessner, 2007a).

Among these barriers is the view that program managers have of the SBIR program as a tax on their development programs rather than as an opportunity to be leveraged. The reasons for this are not clear, although the NRC does suggest that such attitudes may "be an inevitable result" because SBIR funds come primarily from program manager resources. The NRC further states that despite the fact that the "DoD is strongly committed—on paper—to the integration of SBIR into acquisitions," program manager attitudes could perhaps prevent small businesses from being "as fully engaged in the work of defense acquisitions as they might be" (Wessner, 2007a).

Another impediment to integrating SBIR technologies in DoD acquisition programs that the NRC identified is the "TRL gap," that is, the gap between the technology readiness level of a project at the conclusion of Phase II SBIR work and that needed by an acquisition program.[7] A TRL of 8 is required for effective transition, but SBIR projects typically have a TRL of only 3 to 5 when Phase II is complete.[8] The resulting gap requires time and money to resolve for successful transition. Prime contractors contend that the TRL gap is the greatest impediment to SBIR technology transition (Wessner, 2007a).

The NRC found that DoD acquisition officials are concerned about technical, funding, schedule, and contractual risks associated with maturing SBIR technologies and integrating them into larger acquisition programs. An acquisition official must fund the development of SBIR technology to higher levels of readiness with no guarantee that the development will succeed or be timely enough for program integration. Small businesses participating in the SBIR program are also less stable than the larger prime contractors. The result is that when program managers agree to take on SBIR transitions, they assume a number of small, higher-risk contracts that can consume a great deal of their time.

The NRC also identified several impediments confronting small businesses. The lack of a formal SBIR funding mechanism for Phase III can be confusing to such businesses (Wessner, 2007a). Other more systemic issues include

---

[7]   Technology readiness levels range from a low of 1, in which scientific research is beginning to be translated into applied research and development, to a high of 9, in which there is actual application of technology under operational conditions (Defense Acquisition University, 2004).

[8]   See also other discussion in Wessner (2007a) suggesting that the TRL required for successful transition is only 6. The General Accounting Office has also described a TRL of 6 as a "best practice" (General Accounting Office, 1999).

- Timing. The slow acquisition processes make it difficult for small businesses to manage the resulting intermittent cash flow. Similarly, if there is no plan to integrate SBIR technology immediately on completion of the project, the slow pace of acquisition programs may result in a need to market a SBIR technology at the end of Phase II, and result in a significant delay before Phase III can begin.
- Complexity. The DoD acquisition process is complex and very different from commercial business development.
- Roadmapping. Because so much of the technical development in defense acquisitions is planned through "roadmaps," small businesses may find it difficult to integrate into acquisition programs if they cannot place the technologies they are developing early onto the roadmaps.[9]

As small businesses attempt to integrate their SBIR work into DoD acquisition programs, they may encounter still other obstacles, many peculiar to new, small R&D companies. These include

- Finding Phase III funding. If DoD funding for transition from Phase II to Phase III is not immediately available or is inadequate, small businesses may have trouble finding other funding. Venture capitalists may not find defense contracting to be an attractive target for investment for many reasons, including the limited size of the market, long lead times, and "red tape." Small businesses may also not have the resources to pursue venture capital and may be reluctant to relinquish the equity needed to attract it.
- Lack of budget stability. The NRC (Wessner, 2007a) noted two impediments, budget squeeze and contract downsizing,[10] that both can disproportionately affect small businesses lacking a fiscal cushion in pursuing DoD work.

Large prime contractors who do most of the system development for the DoD could help small businesses in SBIR technology transition. Yet, as the NRC has noted, prime contractors may also face impediments to supporting SBIR, including those related to

- Risk. Prime contractors face both technical risk and the risk of doing business with unproven, potentially unstable small subcontractors.
- Timing. Matching the availability of ready technology with the acquisition program schedule can be very difficult.
- Incentives. Small-business partnerships and technology sourcing lack market incentives.

---

[9]   Garcia and Bray (1997) define technology roadmapping as

"a needs-driven technology planning process to help identify, select, and develop technology alternatives to satisfy a set of product needs. It brings together a team of experts to develop a framework for organizing and presenting the critical technology-planning information to make the appropriate technology investment decisions and to leverage those investments.

   Given a set of needs, the technology roadmapping process provides a way to develop, organize, and present information about the critical system requirements and performance targets that must be satisfied by certain time frames. It also identifies technologies that need to be developed to meet those targets. Finally, it provides the information needed to make trade-offs among different technology alternatives."

[10]   Contract downsizing refers to the ability of the government to reduce contracts at its convenience. Budget squeeze refers to the relative inability of small businesses to influence government budget priorities.

Prime contractors also find it difficult to identify the right SBIR technologies and companies. One in particular noted that small businesses rarely approach primes to explore developing partnerships (Wessner, 2007a). The DoD is addressing this particular challenge by hosting an annual showcase event, titled "Beyond SBIR Phase II: Advancing Technological Innovation," to increase stakeholder awareness of SBIR-funded technologies to facilitate partnerships and commercialization.

### RAND Evaluations

RAND researchers have identified similar impediments to SBIR technology transition. Like NRC researchers, Held et al. (2006) found that many DoD program managers view the SBIR program as a tax on their program rather than as an opportunity. That attitude, along with the effort required to manage many small contracts, make the SBIR program "just another mandate to put up with" (Held et al., 2006, p. 74).

The RAND evaluation also found that the bulk of R&D conducted in the program focuses on basic and applied research, leading to immature technology at the end of Phase II. RAND researchers noted that small businesses were often not well equipped to manage subsequent, longer-term technology development because of the cash flow requirements. In sum, a SBIR focus on early stage R&D is not conducive to commercialization or to getting technology rapidly into the hands of warfighters.

RAND researchers also implied that funding in the current program structure is inadequate for full technology development. As a result, they recommended that the DoD

- develop funding structures using SBIR and other R&D funds, such as ManTech, which focuses on the needs of weapon system programs for affordable, low-risk development and production, and Army Venture Capital Initiative Funding, which currently focuses on a program for technologies meeting power and energy requirements for soldier equipment, to ensure a stream of funding for taking a technology through successive stages of development
- package several SBIR projects together to create longer-term, product-focused programs that rely on SBIR as a funding source.

Because prime contractors are the "technology transitioners" for most DoD acquisition programs, RAND researchers suggested that they be part of the process for identifying technology requirements that could be addressed by the SBIR program and that mechanisms for connecting small businesses and prime contractors be added or strengthened. Such steps could ease SBIR technology transition.

### Small Business Technology Council Evaluation

In addition to the independent analyses noted above, the Small Business Technology Council (SBTC) has offered its own assessment on improving technology transfer (Baker, 2007). Like the NRC and RAND, the SBTC identified the technical immaturity of SBIR technologies as a principal reason for transition failure. Unlike the NRC and RAND, which both viewed technological immaturity as a problem rooted in the kind of basic and applied research supported by the program, the SBTC suggested that technology transition could be improved relatively easily through greater program resources and efficiencies.

The SBTC further suggested that DoD acquisition officials may require more incentives to use SBIR technologies. Its research notes

> DoD generally relies on large contractors to make decisions about technology provision, creating a hurdle for small firms. Both Congress and DoD acknowledge that incentives to large firms may help improve contractor performance, and Congress has asked DoD to track incentive use regarding SBIR technologies (Commercialization Pilot Program) (Baker, 2007, p. 4).

Similarly, the SBTC notes a need to make acquisition officials more familiar with the SBIR program. The focus placed on incentives to improve contractor performance and the suggestion that acquisition officials are not familiar enough with the SBIR program suggests that the SBTC believes that at least some of the transition problems occur from ingrained habits.

The SBTC also highlights what it sees as a significant flaw in DoD system development, suggesting that having acquisition personnel take responsibility for maturing technologies before or during integration of systems results in a coupling of technological and integration risks. Because acquisition programs both develop and integrate technologies, it notes, the time required to complete the program is extended, making it more difficult for shorter SBIR technology development cycles to fit into acquisition program cycles.

## Addressing the Impediments to DoD SBIR Technology Transition

The reviews of the DoD SBIR program generally agree on the most important impediments to SBIR technology transition into acquisition programs. These are

- Technical maturity. SBIR research typically results in technology not mature enough for transition to an acquisition program.
- Adequate funding. Funds needed beyond Phase II for development and integration of SBIR technology into a larger system are typically scarce.
- Timing. It is difficult to synchronize SBIR project and acquisition program schedules. Perhaps the biggest timing issue is ensuring that Phase III can begin shortly after the end of Phase II, to provide adequate cash flow for the small business.
- Acquisition culture. Program managers continue to view the SBIR program as a tax and the management required for the program as a burden. They do not view the program as an opportunity to be leveraged.

Program managers are trained very early in their careers that their primary job is the management of cost, schedule, and performance risk. Hence, it is not surprising to see that the SBIR program has impediments related to funding, timing, and technical maturity. The impediment posed by the acquisition culture will be more difficult to address while the perceptions of risk persist and prevent more significant technology transition.

The NRC and RAND studies differ somewhat on how to overcome these impediments. The NRC suggests several best practices to reduce risks resulting from immature technologies. These include returning more of the SBIR "tax" to the program managers for projects that meet their needs, giving program managers more say in SBIR topic selection and drafting of solicitations, and giving program managers more authority to determine which Phase I projects will be selected for Phase II awards. Because program managers are concerned about fielding products, they are likely to use SBIR projects they actually control throughout to solve their near-term technology issues. The NRC argues that this will result in more SBIR projects focusing on later stages of research. RAND researchers' recommendations for reducing technical risk were similar but more explicit, suggesting that the SBIR program should focus less on basic and applied research and more on later-stage research.

To reduce funding risks, the NRC suggests focusing more awards on related topics, thus allowing more research on some problems and improving the probability that a solution will be developed and adopted. The NRC also suggests funding for Phase III activities, giving acquisition program managers incentives to use and integrate SBIR-developed technologies. RAND researchers emphasize reducing funding risk rather than adding resources, recommending reorientation of existing monies such as consolidating SBIR awards around a larger, managed project. RAND researchers also suggest better integration of funding sources such as the Army Venture Capital Initiative and the ManTech Program to create a funding stream large enough to develop technology to full maturity.

To manage timing and schedule risks, both RAND and the NRC studies emphasize program flexibility. Both recognize that SBIR funds are not allocated to particular programs and can be used to address DoD technology requirements as they become known. This is in contrast to funding technology development through more formal budgeting processes that move from DoD planning through the executive branch and ultimately to Congress. The NRC also emphasizes developing tools and practices for greater communication and collaboration between SBIR program managers and acquisition personnel.

Regarding acquisition culture, both the NRC and RAND studies stress that senior leaders need to be involved and to insist on using the SBIR program to develop technologies that can ultimately address warfighter needs. The NRC also recommends improving and increasing the training that acquisition personnel receive regarding the SBIR program to improve transition success.

Finally, the NRC research suggests two reasons for developing better incentives. First, it says that "career oriented" incentives may be required to get more government program managers involved in the SBIR program and Phase II transition. Second, it says that incentives and goals directed at prime contractors would help encourage prime contractor participation in Phase III projects.

## Additional Research Needs

Additional research is needed to best understand all impediments to SBIR transition and how to overcome them. We note in particular four areas for future research that might yield fruitful results.

First, more research is needed on the development and history of SBIR projects. This might include tracking SBIR projects from proposal through Phase II development, Phase III transition, and ultimate commercialization (or its lack).

Second, more research is needed on SBIR companies. This might include identifying the types of companies most likely to participate in SBIRs, what leads them to propose SBIR projects (or dissuades them from doing so), and what characteristics are most likely to lead to success for a SBIR company in technology transition.

Third, more research is needed on incentives for program managers and contractors. The DoD may wish to consider in particular an experimental test of incentives for using the SBIR program, complete with subsequent analysis on which incentives performed best.

Finally, we note that the emphasis on commercialization is resulting in a great deal of experimentation among the SBIR programs run by the military services and the DoD agencies.[11] These experiments need to be carefully monitored to understand what is helping to overcome obstacles to technology transitions and what is not.

---

[11] For example, the Navy has been testing a number of approaches to include technology transition assistance and allocating SBIR topics in accordance with the budgetary contribution of its suborganizations and commands. The Army has been involving its Program Executive Office structure more in the development of SBIR topics. The Air Force now allocates a greater percentage of its topics to its product (acquisition) centers to create an incentive for laboratory activities to seek to partner with them as execution agents for projects addressing identified program office needs. Congress has also authorized the development of a "Commercialization Pilot Program" (Public Law 109-163, Section 252, 2006) that provides more authority and resources for technology transition from the SBIR program to the DoD's acquisition programs.

# CHAPTER SIX

# The Mentor-Protégé Program

## Program Development and Evaluation

In 1986, Congress, concerned about the low participation of small disadvantaged businesses (SDBs) in the DoD's prime contracts, established a goal that 5 percent of prime contract dollars go to such businesses, that is, small businesses that are at least 51 percent owned, controlled, and operated by persons who are socially and economically disadvantaged.[1] DoD did not meet this goal in the late 1980s or early 1990s (Department of Defense Office of Small Business Programs, 2008).

In 1990, Congress therefore enacted Public Law 101-510 establishing the DoD Mentor-Protégé Pilot Program (see *U.S. Code Congressional and Administrative News,* 1990). The program provided incentives for major DoD contractors to help small disadvantaged businesses improve their performance as subcontractors and suppliers on DoD contracts and to increase their participation on other federal and commercial contracts, including in such areas as infrastructure development (e.g., organizational management) and technology transfer. Incentives for mentors included cash reimbursement and credit toward subcontracting goals for small disadvantaged business. Mentors are also able to team with smaller firms that may be developing new technologies on which the mentors could capitalize. Protégés, in turn, would gain experience in and exposure to the DoD marketplace and, ideally, grow to become regular subcontractors or even prime contractors to the DoD or other federal agencies. (For more on this program's background, see General Accounting Office, 1992.)

Initial program implementation was slow. Originally intended to begin in FY 1991, the program had only 16 agreements in place for FY 1992 and 120 agreements in FY 1993 (General Accounting Office, 1994a). In later years, program eligibility was extended to 8(a) firms, firms that employ severely disabled persons, firms in Historically Underutilized Business Zones (HUBZones), and small businesses owned by women or service-disabled veterans (Public Law 106-398, 2000; Public Law 108-375, 2004; Government Accountability Office, 2007).

Data on the performance of the program are limited (General Accounting Office, 1998c). There have been no comprehensive evaluations of program progress, accomplishments, successes, or many other program characteristics. In the late 1990s, the DoD Office of Small Busi-

---

[1]  African Americans, Hispanic Americans, Asian Pacific Americans, Subcontinent Asian Americans, and Native Americans are presumed to qualify. Other individuals can qualify if they show by a "preponderance of the evidence" that they are disadvantaged. All such individuals must have a net worth of less than $750,000, excluding the equity of their business and primary residence (Small Business Administration, 2008d).

ness Programs (then the Office of Small and Disadvantaged Business Utilization) conducted a survey of program participants, but the General Accounting Office faulted this survey for what it considered low response rates—about 60 percent for mentors and 40 percent for protégés. The General Accounting Office also found that many reports on agreements made to the Defense Contract Management Agency (DCMA) were missing information or did not report all required areas. Since these findings, the DCMA has established a separate Mentor-Protégé Program Office with a team devoted to the administration of credit agreements as well as to performing annual reviews of all active reimbursable agreements (Howell, 2008).

In the 1990s, subcontract awards to small disadvantaged businesses, which the program had sought to boost, surpassed 5 percent of subcontracting dollars for both the federal government generally and the DoD specifically. Subcontracting dollars to small disadvantaged businesses for the DoD have been below 5 percent in recent years (Figure 6.1). Noting that the DoD continued to lack data to assess the success of the program, the General Accounting Office concluded that it was unable to determine the relationship between the program and the government's success in meeting the 5 percent subcontracting goal for small disadvantaged businesses or whether the program was responsible for the benefits that participants claimed to have received from it (General Accounting Office, 2001).

A later survey of 48 former protégés who left the program in FY 2004 and FY 2005 did find some levels of satisfaction with the program (Government Accountability Office, 2007). Common industries among respondents included engineering services, computer services,

**Figure 6.1**
**DoD Use of Small Disadvantaged Businesses for Subcontracts, FY 1989 to FY 2006**

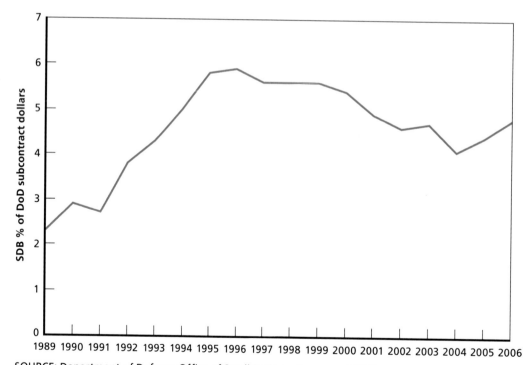

SOURCE: Department of Defense Office of Small Business Programs (2008).
NOTES: The figure likely underestimates the SDB share of DoD subcontracts. Subcontracting data are currently gathered only for contracts worth at least $550,000 (or $1 million for construction projects).
RAND TR601-6.1

and facilities support services. Most protégés reported that the program helped their business development and enhanced their engineering or technical expertise. Many also reported that the program helped them best meet International Organization for Standardization quality standards or Capability Maturity Model Integration certifications. Among the most common areas of support in developing business skills that protégés reported were marketing, organization planning, and contract management. The most common area of engineering or technical support that protégés reported was in quality management programs. Overall, 18 of the 48 protégés reported that they were "very satisfied" with the program, and 16 said that they were "generally satisfied," but two said that they were "generally dissatisfied," and five said that they were "very dissatisfied." The survey response rate was only 63 percent, and no comparable data for nonparticipants are available.

The DoD has also collected annual data from program participants. The Government Accountability Office (2007) questioned the accuracy of many of these data, and comparable data on nonparticipants or on gains for participants directly attributable to the program are not available. Nevertheless, in the nine most recent years for which data are available, protégés have reported aggregate annual net employment gains roughly between 1,400 and 4,000 workers and aggregate annual net revenue gains (as measured in FY 2009 constant dollars) between $190 million and $1.1 billion. Not surprisingly, these trends parallel each other (Figure 6.2). They also appear to be loosely related to trends in the number of agreements in effect each year. Furthermore, the DCMA has also sought to improve the process of data reporting on the

**Figure 6.2**
**Aggregate Employment and Revenue Gains for Protégé Participants, FY 1998 to FY 2006**

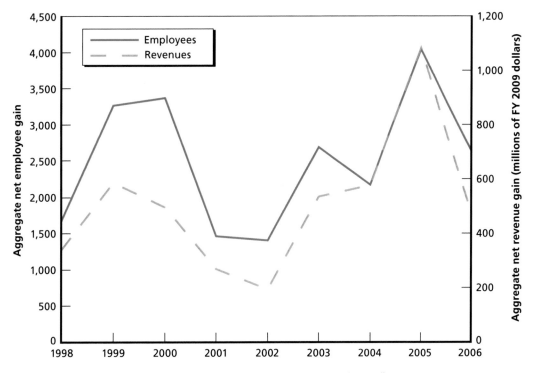

SOURCE: Department of Defense Office of Small Business Programs (annual).

RAND *TR601-6.2*

program by including validation of the data before submittal of the annual report by the Office of Small Business Programs to Congress (Howell, 2008).

As the cumulative number of program participants has grown, so, too, has the share of program participants in DoD contracting opportunities. In FY 2006, protégé firms that had participated in the program since its inception accounted for 1.47 percent of DoD prime contract awards, 4.42 percent of DoD prime contract awards to small businesses, and 10.61 percent of DoD prime contract awards to small disadvantaged businesses. Program participation has become more diverse as well; although small disadvantaged businesses, the traditional focus of the program, constitute more than 90 percent of historical participants, women-owned small businesses constitute more than 5 percent, and those in HUBZones or owned by service-disabled veterans constitute more than 1 percent each.

In recent years, the number of Mentor-Protégé agreements has fluctuated between 142 and 243 (Figure 6.3). Most agreements have been for reimbursement of mentor costs rather than for subcontracting goal credit.

The composition of reimbursement agreements has changed over time (Figure 6.4). In 2000, the Army accounted for just over one in five reimbursement agreements. By 2006, it accounted for nearly two in five. The proportion of such agreements in the Navy and Air Force decreased during this time, whereas those with other DoD agencies increased.

**Figure 6.3**
**Mentor-Protégé Agreements, by Type, FY 2000 to FY 2006**

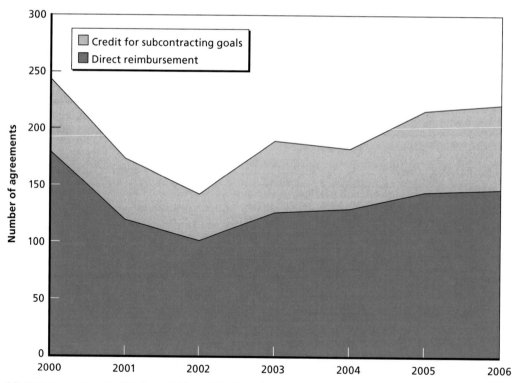

SOURCE: Department of Defense Office of Small Business Programs (annual).
RAND TR601-6.3

**Figure 6.4**
**Mentor-Protégé Reimbursement Agreements, by DoD Component, FY 2000 to FY 2006**

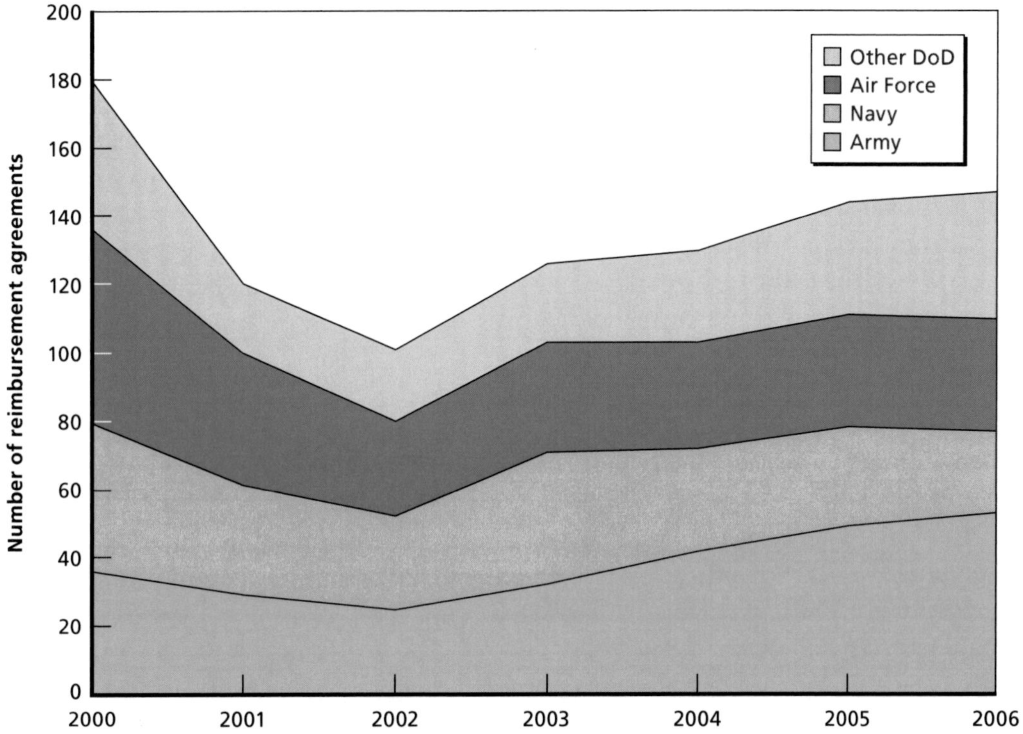

SOURCE: Department of Defense Office of Small Business Programs (annual).
RAND TR601-6.4

The annual appropriation for the program has varied between $18 million and $26 million (Government Accountability Office, 2007). Most funding has been for mentors, including reimbursements for the developmental assistance they have provided to protégés. Other appropriated funds have covered incidental costs such as travel for protégés as well as program office operational expenses.

Reimbursements are key to attracting program mentors. Two surveys have found that without them, the program would be much smaller.

In the first, more than two in three mentors said that they would withdraw from the program if funding for reimbursable agreements were eliminated, and nearly all those that said they would remain indicated that they would reduce their participation without reimbursement (Jennings et al., 2000). This led analysts to conclude that eliminating reimbursement agreements could reduce employment and revenue gains that protégés receive by more than 90 percent (although such a contention appears to depend on all employment and revenue gains that protégés report are indeed attributable to the program).

In the second survey, more than two in three mentors again said that they would not participate in the program if reimbursement agreements were eliminated (Jennings, Koch, and Mercer, 2006). Analysts noted that a program relying on credit agreements would be insufficient because of the technical requirements for fulfilling subcontracting requirements and that procurement preferences would lead to "hollow" agreements made primarily not for the benefit of protégés but for gaining the preferences.

## DoD Perspectives

DoD representatives we interviewed cited three areas of success for the program. These include success stories of individual protégés as recognized by the Nunn-Perry Awards for outstanding mentor-protégé teams, development of protégés in rapidly evolving fields such as robotics, and development of networks that mentors offer protégés beyond the DoD.

The Nunn-Perry Awards were first awarded in 1995 to recognize leading mentor-protégé teams. A DoD representative told us that the program received about 30 nominations annually and that recipients believe that the award is helpful in attracting attention and, ultimately, in funding their products. Applications are reviewed and numerically scored by a panel appointed by the director of the Office of Small Business Programs. The number of awards is set at the discretion of the DoD, with the number of awards to be given, usually eight to 12, determining the scoring cutoff.

Recent award recipients have included partnerships involving

- an environmental services firm that developed software for analysis of hazardous materials sites
- an electronics firm that developed rugged display systems for submarines
- a computer services firm that developed new global positioning system equipment
- an electronics firm that received assistance to implement continuous improvement in its manufacturing and business processes
- an electronics firm that received mentoring in improving its manufacturing and supply chain processes for producing critical batteries
- a computing services and security firm that received mentoring in preparing a strategic business plan
- a communications engineering firm that supported DoD biometrics analysis (Department of Defense Office of Small Business Programs, 2007).

More generally, a DoD representative told us, the program has been helpful in developing and transferring new technologies to the DoD in rapidly evolving industries such as robotics and global positioning system software. This same representative noted that companies resisting participation in the SBIR program because of its funding and other bureaucratic requirements may see more value in the Mentor-Protégé Program. Others noted successful innovations in construction, management, and aerospace industries, as well as the success of the program in developing joint applications for use across the services and other defense agencies.

DoD representatives we interviewed also noted the benefit of the program in exposing protégés to other federal agencies. One, citing cases of mentors helping protégés develop business with NASA and the Department of Homeland Security, noted that the mentors typically have broader and more relevant networks for protégés than the DoD can offer.

Asked about possible program improvements, each DoD representative we interviewed about the program mentioned funding needs. One noted that the expansion of the program to include other businesses besides small disadvantaged businesses has created more competition for program funds, which have been fairly static over the years. This same representative noted that industries in which the program is now likely to be most effective require far more funding than previous projects have needed. Others also noted that the program, lacking a specific earmark or line item of its own, often sees its funds raided by others.

## Conclusions and Recommendations

Although the Mentor-Protégé Program appears to be popular among both mentors and protégés in the DoD, and perhaps has helped boost the size of protégé firms, there is no direct evidence linking it to its goals of boosting participation by small disadvantaged businesses in DoD contracting or to the growth participants have reported. Indeed, in recent years, participation by disadvantaged businesses in DoD subcontracts has been below the statutory goal of 5 percent. Further research on enhancing small-business opportunities through the Mentor-Protégé Program might explore more direct mechanisms between the program and these achievements, as well as how achievements vary by industry or sponsoring DoD component. Such research might, for example, compare participants with other small disadvantaged businesses and compare trends in growth and employment in both, or, indeed, document the numbers of and differences between the firms successfully completing Mentor-Protégé agreements and those that do not.

Perhaps the clearest conclusion from prior research on this topic relates to funding for mentors. Most program agreements are for reimbursement of costs and most mentors receiving such reimbursement say that they would not participate in the program without it. Similarly, DoD representatives told us of continuing funding needs given the greater number of businesses now eligible for the program as well as the expenses that partnerships in some industries may incur. Although Congress may wish to better understand exactly how the program can contribute to the goals set for it before changing its funding levels, it appears that success or even the existence of the program will depend on funding rather than on other credits for mentors.

# CHAPTER SEVEN

# The Effect of Electronic Payment Systems on Small Businesses

Electronic commerce (e-commerce) systems process vendor invoices and also receive documents for DoD goods and services, link contract and funding documents to invoices, and make payments for day-to-day goods and services through paperless electronic means. The Defense Finance and Accounting Service (DFAS), under the direction, authority, and control of the DoD Comptroller, manages the accounting and disbursing mission, making payment to all businesses providing goods and services to the DoD, annually processing more than ten million invoices for more than $300 billion (Defense Finance and Accounting Service, 2008).

How well small businesses can adjust to such systems has been questioned by some, including Congress in authorizing this research. A 2007 poll of small businesses found that more than one in three small-business owners (especially older ones) said that they were not comfortable with information technology issues and opportunities, and only about one in six had Web sites for e-commerce (El Tarabishy, 2007).

In this chapter, we describe the e-commerce systems used by the DoD, review the available literature on e-commerce, and examine the effect of these systems on small businesses. We conclude with recommendations regarding best practices and e-commerce.

## An Overview of DFAS Electronic Commerce Systems

In the traditional DoD payment model, three documents are required to make a payment: the contract, the receiving report, and the invoice. These paper documents usually arrive at the payment office separately and are processed individually as they arrive. Information on these documents is manually entered into the appropriate payment system.

The DFAS is expanding its e-commerce capabilities to improve the accuracy and timeliness of transaction processing, replacing paper-based methods with electronic processing. Its e-commerce objectives are to improve customer service and reduce the cost of operations by providing single-source entry (data entered into one system at one time). This will enable electronic data transmission, electronic verification and payment, and electronic storage of data (Defense Finance and Accounting Service, 2006a).

Table 7.1 lists the primary e-commerce systems that small businesses use to interface with the DoD (Defense Finance and Accounting Service, 2006b, 2007, 2008). Small businesses are required to use some systems to submit invoices and be paid electronically. The optional systems can improve the interface with DFAS and provide contractors greater access to

**Table 7.1**
**DoD Financial Management and Related Systems**

| System | Vendor Use Required or Optional? | System Sponsor |
| --- | --- | --- |
| Dun & Bradstreet Data Universal Numbering System | Required | Dun & Bradstreet |
| Central Contractor Registration and CAGE Code | Required | Office of Federal Procurement Policy |
| Wide Area Workflow | Required | DFAS |
| myInvoice | Required | DFAS |
| Electronic Document Access | Optional | DoD Business Transformation Agency |
| Vendor Pay | Optional | DFAS |
| Web Invoicing System | Optional | DFAS |
| Electronic File Room | Optional | DFAS |
| Electronic Data Interchange | Optional | DFAS |
| Governmentwide Commercial Purchase Card | Optional | General Services Administration |

electronic transaction information. These provide a variety of systems for small businesses to use in working with the DoD.

We describe below each of these systems that small businesses and DFAS use to conduct e-commerce. Unless otherwise noted, the information here summarizes multiple DFAS sources including the *Contract/Pay Information Handbook*, the DFAS Contract/Vendor Pay Web site, the *Wide Area Workflow Getting Started Guide*, and the *Electronic Data Interchange Guide* (see the Bibliography for complete citations).

### Data Universal Numbering System

Dun & Bradstreet (D&B) provides a Data Universal Numbering System (DUNS) number, a unique nine-digit identification code, for each physical location of a business (Dun & Bradstreet, 2008). The DUNS number, which may be acquired at no cost, is required to register with the federal government for contracts or grants. More than 100 million businesses worldwide have been assigned DUNS numbers. The DUNS number links a vendor's physical and mailing addresses, principal names, financial information, industrial classification codes, and socioeconomic status (e.g., whether firm is considered small or "disadvantaged"), among other data elements. It was adopted as the standard business identifier for federal electronic commerce in 1994 and was incorporated into the Federal Acquisition Regulation in 1998 as the federal government's primary contractor identification code for all procurement-related activities. No vendor can perform a DoD contract without a DUNS number. This is a small business's first requirement before pursuing a DoD contract.

### Central Contractor Registration

The Central Contractor Registration (CCR) is the primary vendor registrant database for the federal government. The CCR collects, validates, stores, and disseminates basic vendor data in support of federal government acquisitions. Currently, more than 450,000 active firms are registered in the CCR (Central Contractor Registration, 2008a). After receiving a DUNS number, all DoD vendors must register in the CCR to be awarded a contract, the second step a small business must take to obtain a federal contract. DFAS payment officers use the bank routing data entered by vendors in the CCR to make electronic funds transfer payments, the required means for paying federal contracts since 1996.

### CAGE Code

A Commercial and Government Entity (CAGE) code is a five-digit code assigned to firms by the Defense Logistics Information Service (DLIS), which identifies companies doing business with the federal government (Defense Logistics Agency, 2008). The CAGE code is a product of an 80-column card format, and some legacy logistics and financial systems still use CAGE codes. Therefore, the DoD needs to maintain them in parallel with the DUNS number. Registrants in the CCR are automatically assigned a CAGE code by DLIS.

### Wide Area Workflow

Wide Area Workflow (WAWF) is a DoD-wide paperless invoicing, receiving, and accounting application designed to eliminate paper from the receipts and acceptance process. Its goal is to give defense contractors and DoD personnel the ability to create and transmit invoices, receive reports, access contract-related documents electronically, and facilitate electronic payments by DFAS. The traditional paper method required that data be manually entered into payment systems. WAWF permits the sharing of electronic documents, eliminating paper and redundant data entry and increasing accuracy while reducing the risk of losing a document. Using WAWF, contractors invoice the DoD electronically and DoD personnel receive notification electronically of pending actions. Vendors also benefit from being able to electronically submit invoices, reducing the risk of lost or misplaced documents for them as well, and having online access to contract payment records. WAWF requires digital signatures and certifications, secure socket layer connections, and encrypted data transmissions. With a DUNS number, CCR registration, and CAGE code, a small business can gain access to WAWF, the third essential step toward performing a DoD contract.

### myInvoice

myInvoice is a Web-based application developed for vendors and DoD employees to obtain invoice status. An interactive system, it provides information on invoices submitted and processed against DoD contracts paid by DFAS. myInvoice consolidates invoice data from numerous DFAS payment systems. Vendors can also use myInvoice to determine when payment is scheduled or if something is lacking for payment processing. Obtaining a myInvoice account is the fourth step businesses must take for DoD e-commerce.

### Electronic Document Access

Electronic Document Access (EDA) is a DoD electronic file cabinet for storing and retrieving documents used by vendors and for other DoD activities. It is intended to replace the paper version of contract documents. The primary function of EDA is to allow users access to official DoD documents via a Web browser. EDA serves as a centralized document repository that is used to store contracts and contract modifications, government bills of lading, pay vouchers, and other payment-related documents. EDA also manages the automated workflow of Contract Deficiency Reports, enabling faster and more efficient resolution of contract discrepancies and reducing payment delays. EDA is integrated with WAWF: WAWF is the software interface with the user and EDA is the electronic depository for files. WAWF users, including vendors, have access to contract documents that match their validated DUNS or CAGE codes. Although optional for small businesses, EDA is the final vendor registration a small firm needs to fully engage in e-commerce (Business Transformation Agency, 2008a). Without EDA access, vendors cannot view the supporting documents that DFAS uses to process payments.

### Vendor Pay

Vendor Pay processes payments for common goods and services contracts not administered by the Defense Contract Management Agency plus other miscellaneous payments. It operates from multiple DFAS offices. Vendors interact with Vendor Pay, which interfaces electronically with 16 systems to make payments in support of the Army, Navy, Air Force, Marine Corps, and defense agencies. Some of these payment systems support only a single commodity type, such as fuel or rations. Others are used at multiple locations for a variety of goods and services. Small businesses may require Vendor Pay access depending on the goods or services they provide and the DFAS office that processes their payments.

### Web Invoicing System

The DFAS Web Invoicing System (WInS) is a Web-based application that allows current paper-based vendors to send invoices electronically. WInS converts the vendor's invoice into Electronic Data Interchange (EDI) format understood by the DFAS electronic payment systems. Edits and validations occur before the file is converted and forwarded through the DoD information processing infrastructure to the payment system. An invoice log allows the vendor to review, edit, and print invoices for transmission tracking and archive purposes. DFAS WInS processes more than 100,000 invoices per month totaling more than $3 billion (Defense Finance and Accounting Service, 2008). Small firms using paper invoicing can use WInS as a way to forgo more complex e-commerce systems.

### Electronic Data Interchange

EDI is the computer-to-computer exchange of routine business information in a standard format. DFAS has implemented EDI transactions to support commercial pay and accounting processes. EDI transactions eliminate the need to reenter critical data in commercial pay systems and accounting systems. EDI invoicing capability, coupled with increased use of electronic funds transfer, has helped decrease errors in and improve processing of payments. DFAS Commercial Pay uses EDI to submit commercial invoices and other financial transactions for multiple systems including the Improved Awards Processing System, the Mechanization of Contract Administration Services system, and the Standard Automated Materiel Management System. Firms can establish EDI with DFAS systems to automate routine processes.

### Electronic File Room

The Electronic File Room (EFR) is a Web-based, view-only portal for the Electronic Document Management system of DFAS. The EFR allows government and vendor access to official DoD documents stored on the local contract pay and vendor pay databases. It reduces delays in accessing documents and provides a single source for electronic search and retrieval of documents as well as increased visibility of all procurement and payment actions.

### Governmentwide Commercial Purchase Card

The Governmentwide Commercial Purchase Card (GCPC) is used for more than eight million purchases annually exceeding $6 billion. The General Services Administration introduced it in 1993 to reduce total costs and simplify and speed small transactions. The DoD adopted it in the mid-1990s. Using a GCPC, DoD employees make purchases of up to $25,000 from small and large businesses or pay contractors. Because most of these transactions are for less than $3,000, DoD users are not required to report them in the FPDS-NG. Each employee with a GCPC has an individual account similar to a personal charge card. The card-issuing banks[1] provide direct payments to the vendors for DoD GCPC purchases. Monthly, DFAS makes payment in full to the bank holding the GCPC account. The cardholder, DFAS, and the banks do not capture transaction data regarding business size (Johnson, 2007).

## The Path to E-Commerce

Within two weeks, a firm can be ready to do business with the DoD (Table 7.2). Information for completing this process is readily available at the Small Business Administration Web site and other Web sites such as those managed by the DoD Office of Small Business Programs, the Central Contractor Registration, and area Procurement Technical Assistance Centers. Each registration requires logins, and the passwords and password conventions differ. A newer computer with a high-speed Internet connection can effectively conduct e-commerce with DFAS; a dedicated computer with exclusive software is not required (Johnson, 2007).

**Table 7.2**
**Process to Become E-Commerce Ready**

| Action | Time to Complete and Process | Web Site/Reference |
| --- | --- | --- |
| 1. Obtain an employer identification number or taxpayer identification number online | 1 to 3 days | https://sa2.www4.irs.gov/modiein/individual/index.jsp |
| 2. Determine applicable industrial classification codes | 0.5 days | www.osha.gov/oshstats/sicser.html; www.dlis.dla.mil/h2; www.census.gov/epcd/naics07; www.fpds-ng.com/downloads/psc_data_10242006.xls |
| 3. Obtain a DUNS number | 1 to 3 days | http://fedgov.dnb.com/webform/displayHomePage.do |
| 4. Request a CAGE code | 2 days | www.ccr.gov |
| 5. Register and configure computer for WAWF | 1 day | https://wawf.eb.mil |
| 6. Register with myInvoice | 0.5 days | https://myInvoice.csd.disa.mil |
| 7. Register with EDA | 5 days | http://eda.ogden.disa.mil |

---

[1]  These banks—also known as GSA SmartPay Vendors—are US Bank, Citibank, First National Bank of Chicago, Mellon Bank, and NationsBank.

The Web site for the Wide Area Workflow system documents its system configuration requirements (Business Transformation Agency, 2008b; Defense Finance and Accounting Service, 2007). The myInvoice system requires the Sun Java™ plug-in, which can be downloaded free from the Web, and EDA requires the Adobe Acrobat Reader™, which can also be downloaded free. DFAS e-commerce systems work best with Windows-based applications including Internet Explorer 7.0™. Vendors can invoice the DoD electronically using a file transfer protocol or EDI, which requires a test of initial transactions with the Joint Interoperability Test Command.

DoD vendors are required by law to submit invoices electronically (Public Law 106-398, 2000). They may request a waiver under limited circumstances, and some do resist e-commerce. Nevertheless, according to DFAS, most vendors prefer the timelier processing of electronic transactions (Coulter, 2008). Small businesses still using paper invoices can use the WInS to submit invoices electronically from any Internet-ready computer, including, for example, public-use computers at local libraries.

System training is available online and can be completed in a few hours (Brewin, 2007). Web-based training is available for WAWF, including an area where vendors can practice submitting invoices.[2] The DFAS Web site also provides additional e-commerce support to users, including a page on frequently asked questions. All these resources are available at no cost to the user.

## DoD E-Commerce in Perspective

Outside the DoD, competitive pressures and the cost of implementation appear to be key determinants in the e-commerce decisions of a firm. The owner's view of e-commerce and its benefits is also important; a positive attitude toward e-commerce and a belief that it will enhance profitability make owners more likely to adopt electronic business procedures, with firms perceiving e-commerce positively tending to credit greater benefits to its implementation (Quaddus and Hofmeyer, 2007; Al-Qirim and Corbitt, 2004; Elia, Lefebvre, and Lefebvre, 2004; Cho, 2006; and Fisher and Craig, 2005). Outside the DoD, cost is also cited as a determinant for firms considering e-commerce. Below, we review literature on DoD, other government, and commercial adoption of e-commerce and what that implies for the efforts the DoD now expects of its suppliers.

### Government Accountability Office Research
In a review of FY 2004 payments, the Government Accountability Office (2006d) found that DFAS paid late 10 percent of all invoices but 14.5 percent of small-business ones. The report attributed this to the priority given to complex invoices requiring more time to process and to those for greater amounts that would require interest payments if paid late, with such invoices more likely to come from large firms than small ones. The Government Accountability Office noted that although the DoD reported reducing the frequency of late payments, these improvements came through dedicating additional resources to resolve payment problems rather than correcting system weaknesses.

---

[2]   WAWF Web-Based Training (2008).

The Government Accountability Office also found that delayed processing of payment documents at the time was caused in large part by extensive DFAS paperwork and lack of integration between DoD payment, accounting, and logistics systems. It noted the fielding of WAWF could facilitate electronic exchange of payment data and documents, but it cautioned that the program needed focused management and performance metrics.

The research also highlighted the problems of 17 small-business owners who had been paid late multiple times. Many ultimately had to use personal resources or credit to finance daily operations, and three said they were concerned about firm survival. Although the problems with these owners appeared to be real and significant, the fact that DFAS processes nine million invoices annually suggests that such problems might not have been representative of all small businesses and possibly have since been corrected by improved DoD systems.

The Government Accountability Office (2006d) offered several recommendations for improving DoD processes, suggesting that it

> (1) clarify its management structure for WAWF and provide strategic direction for DOD and the military services in their efforts to implement and effectively utilize WAWF; (2) establish a strategic plan that defines the roles and responsibilities of the various organizations that are integral to the program's success; (3) develop performance metrics to measure the success of the program; (4) consider incorporating, as part of the WAWF application, a data element that would flag invoices submitted by small disadvantaged business so that they could be paid early, in accordance with DOD policy; and (5) require the military services and defense agencies to process all DOD receiving and acceptance reports and other supporting payment documentation electronically.

The DoD concurred with the recommendations, including one to expand WAWF metrics and consider a data element that would flag invoices submitted by small disadvantaged businesses (Government Accountability Office, 2006d).

### SBA Study on Trends in E-Commerce

A study for the SBA found four major barriers that could prevent small businesses from adopting e-commerce: technology, market conditions, resource constraints (capital and personnel), and regulatory barriers (Innovation & Information Consultants, Inc., 2004). Technology barriers include problems integrating e-commerce with other software and lack of high-speed Internet access. Market condition barriers included weakening economic conditions that may slow the adoption of e-procurement. Resource barriers included the lack of capital small firms had to invest in e-commerce and the personnel to maintain it. Regulatory barriers included the existence of multiple government procurement Web sites, differing formats and procedures throughout the federal government, concerns regarding security and privacy in dealing with e-commerce, and increased use of credit cards for small purchases that many small firms did not have the capability or desire to handle. Yet the study did not find "any significant lag in the actual adoption of e-commerce by small business" when compared to large business (Innovation & Information Consultants, Inc., 2004, p. 5).

The SBA recommended that the federal government move toward a single interface and registration point for small businesses; continue to provide training, support, and networking opportunities for them; and train federal employees in the benefits of electronic commerce. The DoD has moved to address many of these recommendations. The Defense Supplement to the Federal Acquisition Regulation standardizes procurement processes across the DoD. The DoD uses the federal government's interface, FedBizOpps, to post solicitations of more than $25,000; those for smaller amounts are posted on DoD contracting-office Web sites or other electronic means, as well as being advertised by manual notification methods including bulletin boards, mailings to bidder's lists, and direct contact with local vendors via fax or email, giving small businesses ready and appropriate access to these opportunities.

### E-Commerce in the Commercial Sector

Electronic commerce is widely used by large commercial firms to manage transactions with suppliers. Wal-Mart Stores, Inc., for example, has an extensive e-commerce program for its suppliers. Wal-Mart suppliers must develop EDI capability and use Wal-Mart's Retail Link® Web-based supply management system to download purchase orders and track shipments. Firms must register with Dun & Bradstreet at their own expense before seeking to be a Wal-Mart supplier (Wal-Mart Stores, Inc., 2008).

Cessna, Honeywell, General Dynamics, Hewlett Packard, and Northrop Grumman all maintain e-commerce systems including procurement, delivery, quality, and invoicing. Like the DoD, some firms have multiple systems. Northrop Grumman's Online Automated Supplier Information System links a supplier to as many as six supplier management Web-based applications (Northrop Grumman Corporation, 2008). Cessna and Honeywell use the Web-based Harmony Order Management™ system, developed by ESIS, Inc., and augmented at Cessna by Ariba procurement management software, to conduct e-commerce with thousands of suppliers, including forecasting, ordering, tracking, and shipping functions (ESIS, 2005, 2008). General Dynamics Land Systems implemented an Oracle-based enterprise requirements planning system in January 2008 to support electronic payment orders and requests for quotation, interactive delivery order management, and processing of receiving, inspection, invoicing, and payment information (General Dynamics Land Systems, 2007). Hewlett Packard uses a third-party electronic invoicing system that larger vendors are required to use to receive payment (Hewlett Packard, 2008). In sum, similar to the DoD, commercial firms are mandating that Web-based systems be used by suppliers but at minimal or no cost to suppliers rather than requiring that suppliers purchase expensive e-commerce software packages. Also like the DoD, large firms often have multiple e-commerce solutions for their suppliers.

### DoD Data on E-Commerce

DoD data on e-commerce with small businesses are limited. DFAS Columbus operates the Contract Pay system, which pays large-dollar, long-term, complex contracts, such as those for weapon systems. Contract Pay makes payments to more than 16,000 contracts on more than 317,000 contracts annually (Defense Finance and Accounting Service, 2008). In contrast, the Vendor Pay system processes more than three-quarters of all electronic payments and is the primary payment system for contracts with small businesses. Between FY 2005 and FY 2007, DFAS processed more than nine million Vendor Pay invoices and made approximately 14 million separate payments each year (Coulter, 2008).

DFAS representatives were very cooperative in providing information to us, but the data are scant. DFAS e-commerce systems are not programmed to track invoices, contract payments, payment delays, or interest penalties by business size or category. DFAS provided data on helpdesk calls but could not distinguish the size of businesses, preventing us from determining whether small businesses have more trouble than larger businesses with these systems. In addition to variables by business size, future research may wish to consider extracting relevant data on DFAS training and access to the e-commerce systems.

The data DFAS did provide indicated that it processed more than $200 billion of invoices using WAWF and Vendor Pay in FY 2007. It rejected 3.7 percent of invoices submitted for payment. Between FY 2007 and FY 2007, DFAS paid about $30 million in interest each year on late DoD payments, with about $5 million of these payments resulting from contracting officer representatives not submitting timely receiving reports to DFAS (Coulter, 2008). Unfortunately, neither data specific to small businesses nor more historical data are available.

### Los Angeles County Government E-Commerce

To assess how other governmental agencies manage e-commerce issues, we spoke with representatives of the Los Angeles County Office of Small Business and Procurement Technical Assistance Center (LA OSB/PTAC). This office provides information and technical assistance to small businesses on procurement opportunities, licensing, permits, certification, and financing at the county, state, and federal levels. It also hosts workshops and training for businesses on selling goods and services to county, state, and federal governments. It is one of 93 Procurement Technical Assistance Centers nationwide funded by the DoD to help small businesses obtain subcontracts with prime defense contractors.

LA OSB/PTAC personnel reported that they had not observed any notable effects of DFAS e-commerce initiatives on small firms trying to enter into business with the DoD. When WAWF was first introduced, these personnel told us, there was an increase in inquiries regarding registration and training, but this has tapered off.

The technology required or the technical nature of e-commerce does not appear to pose a problem for small businesses. The LA OSB/PTAC program director told us "newly established firms typically have no problems, while some older, family-run businesses will designate a 'tech-savvy whiz-kid' to figure out e-commerce. Sometimes it's a son, daughter, or grandchild coming into the business" (Cabreira-Johnson, 2008).

LA OSB/PTAC personnel have seen no indication that DoD e-commerce systems are barriers for small businesses. Rather, they observe that vendors appear to be pleased with e-commerce and electronic payments and rarely report payment problems. The program director notes, "Our experience with vendors doing business or willing to do business with federal agencies shows that small firms will and do adjust to the technology because they have no choice. It does take some time to get acclimated, however" (Cabreira-Johnson, 2008).

### Future DFAS Plans

By the end of FY 2009, DFAS hopes to process 90 percent of payments using e-commerce. To accomplish this, it is planning enhancements to WAWF and improvements to older payment systems. As the military services and defense agencies upgrade their procurement and management systems, DFAS legacy systems will be retired. DFAS personnel believe that invoice submission and processing via e-commerce will improve payment timeliness, decrease human error, and reduce interest charges for late payments (Coulter, 2008).

Contract Deficiency Reports (CDRs), once processed manually using paper DD Form 1716, were replaced in June 2006 with an automated CDR application. This helps DFAS to report, track, and resolve CDRs including those on conflicting quantities or missing lines of accounting. DFAS personnel contend that the automated application reduces the correction time of contract errors and shortens payment delays (Walter, 2008).

## Conclusion

Although e-commerce, e-procurement, and information technology may be perceived as an impediment, the effect of DoD's e-commerce systems on small-business opportunities appears to be minimal. In particular, we found

- There is no clear evidence that DoD e-commerce is a significant barrier to small businesses. Although previous research on e-commerce indicates that limited resources, unfamiliarity with technology, and indifference tend to hinder small-business adoption of e-commerce, there is no empirical data indicating that DoD e-commerce systems such as Vendor Pay, myInvoice, and Wide Area Workflow are barriers to most small businesses. DoD e-commerce systems do not require the significant resources, training, or equipment that most literature reports as barriers for small businesses.
- The software and technology required for DoD e-commerce is neither complex nor costly to obtain. Computers purchased today usually contain the required equipment and software. DFAS has dedicated helpdesks to help resolve WAWF problems. The software required can be downloaded free from the Web. Free training is available from DFAS and other assistance is also available from the SBA, DoD agencies, and Procurement Technical Assistance Centers.
- E-commerce appears to improve the ability of small businesses to invoice the DoD and receive payment. Earlier problems that the Government Accountability Office noted regarding late payments stemmed from a paper-driven system. The Government Accountability Office also noted that the WAWF could reduce these problems through greater system integration. Indeed, many small firms adopting e-commerce prefer it because it eliminates problems.

In sum, although small businesses must clear some hurdles to engage in e-commerce with the DoD, the resources required to do so are modest and extensive support exists to assist small businesses. There is some evidence that small businesses prefer e-commerce in dealing with the DoD, although better data on the use of e-commerce tools or even of helpdesk inquiries by firm size could yield more insights.

# CHAPTER EIGHT

# Prospects for Small-Business "Graduation"

Among the aims of small-business policy are preserving free competitive enterprise and maintaining and strengthening the overall economy of the nation (Public Law 85-536, as amended, 2004). Small firms helped by federal policies in support of these aims may, over time, become larger firms. Yet there may be impediments, both in the broader economy and within the DoD, to such "graduation."

Increasing specialization among defense firms and their suppliers may make it difficult for new firms to develop other, commercial customers (Hayward, 2005). Such specialization may perhaps increase the risks that small firms selling military goods and services face in modifying their goods and services for other public, industrial, or commercial customers. Although small firms specializing in defense goods and services may find little risk in offering somewhat modified products for other government customers, their risks may increase as they offer different products using different technologies to industrial or commercial customers (Hougui et al., 2002).

Further complicating any movement toward graduation from procurement preference programs by some firms are current size standards for such programs that encourage small businesses to "hover around the edges of staying small. Many are fearful of graduating because they become too large for the small-business programs, while remaining too small to compete with the giants in their industry" (Clark and Moutray, 2004). Size standards may also provide disincentives for government buyers and contractors with small-business subcontracting goals to help their small suppliers grow. If a supplier loses its small-business status, for example, a buyer may be forced to find and qualify another small supplier to maintain its ability to meet small-business goals (Hofman, 2006).[1]

Several issues beyond DoD's control can affect small-business graduation rates. For example, many small businesses may choose not to grow. One survey of small businesses found that only about half wanted to grow and less than 10 percent wanted to be "growth" firms (Dennis, 2001).[2] Further, some small businesses, particularly those that have intellectual assets and next-

---

[1] Before July 1, 2007, agencies could continue to count multiyear contracts initially signed with a small business that outgrew its preference thresholds or was acquired by a large company toward meeting their small-business goals for the duration of the contract (Zwahlen, 2007).

[2] The extent of small-business interest in government contracting opportunities is also not clear. A 2004 survey found "winning contracts from federal/state/local governments" to rank 69th out of 75 issues in importance for small-business owners, with issues such as costs for insurance and workers' compensation, fuel, and taxes topping the list (Phillips, 2004). An analysis of the survey results notes

> For years, both the federal and state/local governments have been trying to increase the share of procurement business going to small firms. This survey indicates that despite the bundling of contracts into larger and larger amounts, this has not been a significant problem for small-business owners. . . . Owners generally felt that they did not produce a product or

generation products (i.e., future products derived from a preceding product) are acquired each year by other, often large businesses.[3] Although small-business acquisitions can be rewarding for their owners, they reduce the population of small businesses that might be capable of graduating. Further, acquirers are likely to target small businesses with some of the best prospects for growth.

In this chapter, we explore the graduation of firms from small-business programs and seek to identify any impediments or disincentives to the transition from being a small to becoming a middle-size business, or beyond the SBA thresholds for small businesses. We do so using contract action data reported as awarded to small businesses from the DD350 and FPDS-NG. Although small-business status is based on the status of the parent company, the FPDS-NG reports contract actions by contractors,[4] which are "children" of the parent firm from which small-business status is determined (see the appendix for more on data and interchangeable terms for firms, establishments, and contractors). It does not include data to link a contractor to its parent firm. CCR data do link contractors to their parent DUNS number but provide no additional information beyond the DUNS number. Consequently, our analyses of contracting data and graduation focus on contractors rather than on firms. Put another way, lacking complete data on graduation by firms, we use proxy data on contractors to determine how many firms may be graduating.

As discussed above, the quality of these data varies and can be inconsistent. For example, within one fiscal year, contract actions for the same company and NAICS code can be listed as those for small business on some contracts but not on others. Similarly, the same contract can have actions reporting the same business as both large and small in providing different goods or services within the same industry. Because there are no archives to verify the status of businesses in earlier fiscal years, we counted a business as small if one of its contract actions was reported as small.

We also use CCR data (as of May 1, 2008) on total revenue and current size status by NAICS codes for businesses seeking government contracts. Last, we use Dun & Bradstreet data prepared for the Department of Defense to identify a contractor's parent company and small-business status for the parent's primary NAICS code from 2001 to 2006. We first characterize the nature and stability of the small-business contracting environment within DoD and its possible effect on small-business growth.

---

service that the government wanted to buy. Perhaps this is a problem of perception as much as paperwork requirements. Or small firms wanting to get government business recognize they really do need to band together to be able to produce large quantities at competitive prices (Phillips, 2004, pp. 11–12).

[3]  In 2005 alone, there were 70 mergers of firms that contract with the government in the Washington, D.C., region (McCarthy, 2006). For comparison purposes, we note FY 2004 DD350 data contained almost 7,000 contractors based in the District of Columbia, Maryland, or Virginia.

The acquisition of outside talent and solutions that fit new product and technology niches can deliver quick access to untapped opportunity and help companies move into adjacent markets (DeBeer, 2006). Cisco Systems is well known for this approach, which many other companies have adopted (Byrne, 1998; White and Vara, 2008).

Small firms can be attractive to larger ones for many reasons. For example, after September 11, 2001, security clearances became more difficult to obtain, leading many firms to acquire firms that had them and were hence able to bid on more lucrative classified projects (Fields, 2004; Pae, 2003).

[4]  A contractor, identified by a unique DUNS number (also referred to as Contractor I.D. Code), represents a specific site or separate unit (e.g., a production facility, retail outlet, or distribution facility) of a firm. A parent firm (e.g., Home Depot) may house more than one contractor (e.g., each Home Depot store is a contractor). This is more likely to be the case with very large companies than with small ones.

## Contracting Environment

We began by analyzing the distribution of the number of contracts with at least one action coded as going to small business from FY 1997 to FY 2007,[5] referred to here as small-business contracts, as well as the distribution of contractors with those contracts. Table 8.1 summarizes these data, including key statistics for the time period, number of small-business contracts per contractor, and total dollars per contractor in the time period.

The average number of DoD small-business contracts per contractor with at least one small contract action over the 11 years was 7.3, but the median was two contracts per contractor. That is, most contractors had only one or two contracts. One percent of DoD's contractors received 96 or more contracts (an average of 8.7 contracts or more per year). And one contractor, Kampi Components, a small disadvantaged business, received 9,206 contracts from the DoD (an average of about 837 contracts per year). The sales and administrative costs associated with winning and managing lots of contracts increase a supplier's costs (Hahn, Kim, and Kim, 1986) and may make it less competitive for growth purposes. Mean contract value was $2.73 million; median contract value was $57,827. Such values, spread over an 11-year period, are unlikely to sustain a small business.

These statistics indicate that the distribution of the number of small-business contracts, that is, contracts with at least one action coded as going to a "small" business, per contractor is highly skewed. Table 8.2 shows the distribution of contractors by the number of small-business contracts they held. Columns in the table provide

- number of contracts per contractor: the total number of small-business contracts per contractor over the past 11 years clustered into groups from one contract to 1,000+ contracts received[6]

**Table 8.1**
**Statistics for the Distribution of DoD Small-Business Contracts and Dollars per Contractor, FY 1997 to FY 2007**

| Statistics | Total No. of Contracts/Contractor over 11 Years | Total Dollars/ Contractor over 11 Years |
|---|---|---|
| Max | 9,206 | 69,789,700,000 |
| Average | 7.3 | 2,728,034 |
| Median | 2 | 57,827 |
| Min | 1 | 1 |

SOURCES: FY 1997–2004 DD350 and FY 2005–2007 FPDS-NG data.

NOTE: In Tables 8.1 through 8.7 the following were removed from analysis: Invalid = 2, GENERIC DUNS = 13, ZERO DOLLARS = 835, NEGATIVE DOLLARS ONLY = 1,094

---

[5]  We selected 11 years for our period of analysis for a number of reasons. First, we wanted to include several years before September 11, 2001, because of the possible subsequent effects of that event on DoD contracting levels and practices. Second, we wanted to make the period long enough to reduce the effects of multiyear contracts at the beginning and end of the time period. Third, given the shift from DD350 to FPDS-NG between FY 2004 and FY 2005, we wanted to have an adequate number of years in one data system. We eliminated two contractors with invalid codes, 13 with generic DUNS numbers representing categories of vendors not specific to any individual or entity, and contractors with zero or negative dollars.

[6]  Data are for contracts on which there has been an action in any year from FY 1997 to FY 2007. Contracts may extend more than one year, as we will discuss below.

**Table 8.2**
**Contractors, by Small-Business Contracts Held, FY 1997 to FY 2007**

| No. of Contracts/ Contractor over 11 Years | No. of Contractors | Percentage of Contractors | Cumulative Percentage of Contractors |
|---|---|---|---|
| 1 | 75,945 | 45.98 | 45.98 |
| 2 | 28,188 | 17.07 | 63.05 |
| 3 | 14,722 | 8.91 | 71.96 |
| 4 | 8,941 | 5.41 | 77.37 |
| 5 | 6,053 | 3.66 | 81.04 |
| 6 | 4,462 | 2.70 | 83.74 |
| 7 | 3,271 | 1.98 | 85.72 |
| 8 | 2,569 | 1.56 | 87.27 |
| 9 | 2,033 | 1.23 | 88.51 |
| 10 | 1,731 | 1.05 | 89.55 |
| 11–50 | 13,901 | 8.42 | 97.97 |
| 51–100 | 1,817 | 1.10 | 99.07 |
| 101–500 | 1,385 | 0.84 | 99.91 |
| 501–1,000 | 108 | 0.07 | 99.97 |
| 1,000+ | 44 | 0.03 | 100.00 |

SOURCES: FY 1997–2004 DD350 and FY 2005–2007 FPDS-NG data.

- number of contractors: the number of contractors in each group
- percentage: the percentage of all contractors in the group
- cumulative percentage: the cumulative percentage of contractors in each group.

The table shows that 75,945 contractors (almost 46 percent of all contractors) received only one small-business contract during the 11-year time period.

We next analyze the extent to which small-business contractors, that is, contractors holding at least one contract with at least one action deemed as going to small businesses, span more than one fiscal year. Longer contracts can provide more stability for business growth. We analyzed contract action data from FY 1997 to FY 2007 and identified how many contracts had continuous or noncontinuous actions in one or more years.[7] Table 8.3 presents the results of these analyses. Columns in the table provide

- number of years: the number of fiscal years within which a contract had actions
- number of contracts in consecutive years: the number of contracts with actions appearing in one or more continuous years

---

[7]  Multiyear contracts that overlap the beginning or end of this time segment may lead to an undercount of contracts spanning more than one year, whereas those beginning or ending in midyear may lead to an overcount.

**Table 8.3**
**Distribution of Small-Business Contracts, by Number of Years with Actions, FY 1997 to FY 2007**

| No. of Years | No. of Contracts in Consecutive Years | No. of Contracts in Nonconsecutive Years | Total No. of Contracts | Percentage of Total Contracts | Percentage of Total Contract Dollars |
|---|---|---|---|---|---|
| 1 | 0 | 1,071,802 | 1,071,802 | 90.05 | 19.50 |
| 2 | 62,423 | 6,771 | 69,194 | 5.81 | 13.20 |
| 3 | 20,081 | 3,854 | 23,935 | 2.01 | 27.70 |
| 4 | 10,153 | 2,221 | 12,374 | 1.04 | 10.60 |
| 5 | 5,919 | 1,120 | 7,039 | 0.59 | 10.00 |
| 6 | 3,208 | 407 | 3,615 | 0.30 | 8.40 |
| 7 | 959 | 241 | 1,200 | 0.10 | 4.80 |
| 8 | 430 | 120 | 550 | 0.05 | 2.60 |
| 9 | 403 | 15 | 418 | 0.04 | 2.50 |
| 10 | 27 | 7 | 34 | 0.00 | 0.40 |
| 11 | 8 | 0 | 8 | 0.00 | 0.30 |
| Total | 103,611 | 1,086,558 | 1,190,169 | 100.00 | 100.00 |

SOURCES: FY 1997–2004 DD350 and FY 2005–2007 FPDS-NG data for transactions greater than $2,500–$3,000.
NOTE: Columns may not sum to 100 percent because of rounding.

- number of contracts in nonconsecutive years: the number of contracts with actions in one or more nonconsecutive years
- total number of contracts: the total number of contracts with actions in one or more years—the sum of columns two and three
- total percentage of contracts: the percentage of small-business contracts over the 11-year period
- total percentage of contract dollars: the percentage of total small-business contract dollars awarded over the 11-year period.

Short-term and volume contracts create supplier uncertainty in future demand, increase supplier costs, and limit supplier planning (Hahn, Kim, and Kim, 1986). This can also make it more difficult for small businesses to acquire adequate capital financing and partners to support growth (Baum, 2004). From FY 1997 to FY 2007, the DoD reported 1,190,169 small-business contracts. Ninety percent of these contracts representing almost 20 percent of DoD's total spending with small businesses had contract actions in only one fiscal year. The disparity in length of contracts and total contract dollars highlights the difficulty that DoD small contractors with modest, sporadic business may have in growing.

Because financial stability is one criterion many commercial buyers use to select suppliers (Bowman, 2007), we also assessed small-business contractors by the number of years they held contracts between FY 1997 and FY 2007. Continuous business provides a more stable base from which small businesses can grow. Accordingly, in Table 8.4 we present columns of data on

- number of years: the number of fiscal years within which a small contractor had contract actions
- number of contractors in consecutive years: the number of small contractors receiving actions in consecutive years
- number of contractors in nonconsecutive years: the number of small contractors receiving actions in one or more nonconsecutive fiscal years
- total number of contractors: the total number of small contractors by the number of years in which they received actions
- total percentage of contractors: the percentage of total contractors by the number of years in which they received actions
- total percentage of contractor dollars: the percentage of total small-business contractor dollars awarded over the 11-year period.

Over the past 11 years, 46 percent of small contractors had actions in only one year, receiving less than 2 percent of DoD spending on small-business contracts. Such short-term, low-volume relationships increase supplier uncertainty and costs and do not promote planning critical for investment and growth. More than 36,000 small contractors, or 22 percent, had discontinuous business with the DoD, that is, they had business in at least two nonconsecutive

**Table 8.4**
**Distribution of Small Contractors, by Number of Years with Contract Actions, FY 1997 to FY 2007**

| No. of Years | No. of Contractors in Consecutive Years[a] | No. of Contractors in Nonconsecutive Years[a] | Total No. of Contractors | Total Percentage of Contractors | Total Percentage of Contractor Dollars |
|---|---|---|---|---|---|
| 1 | 0 | 75,329 | 75,329 | 45.61 | 1.90 |
| 2 | 20,334 | 10,392 | 30,726 | 18.60 | 3.10 |
| 3 | 11,544 | 7,440 | 18,984 | 11.49 | 3.90 |
| 4 | 6,535 | 5,540 | 12,075 | 7.31 | 4.20 |
| 5 | 4,538 | 3,936 | 8,474 | 5.13 | 5.50 |
| 6 | 3,000 | 2,731 | 5,731 | 3.47 | 5.60 |
| 7 | 1,614 | 2,173 | 3,787 | 2.29 | 5.70 |
| 8 | 1,113 | 1,696 | 2,809 | 1.70 | 5.60 |
| 9 | 1,003 | 1,232 | 2,236 | 1.35 | 22.50 |
| 10 | 851 | 1,056 | 1,907 | 1.15 | 9.50 |
| 11 | 3,112 | 0 | 3,112 | 1.88 | 32.50 |
| Total | 53,644 | 111,526 | 165,170 | 100.00 | 100.00 |

SOURCES: FY 1997–2004 DoD-wide DD350 data for transactions greater than $25,000; FY 2005–2007 FPDS-NG data for transactions greater than $2,500–$3,000.

NOTE: Columns may not sum to 100 percent because of rounding.

[a] Unique company I.D. codes.

years. Discontinuous business also increases supplier uncertainty, ultimately increasing supplier costs, making suppliers less competitive and less likely to grow (Hahn, Kim, and Kim, 1986).

Less than 2 percent of DoD's small contractors had actions in all 11 years. These 3,112 contractors received more than 40 percent of the DoD's total spending with small businesses. In other words, DoD small-business spend is concentrated among a relatively small number of contractors. These small contractors, with larger, more stable DoD spending, are more likely to have the wherewithal to grow.

## Small-Business Graduation

It is challenging to use current data to identify businesses that have graduated from small-business programs. Data on business revenue, although collected by the CCR, are not archived annually. Hence, we cannot track changes in small-business revenue over time. We can only verify the most recently reported size and average annual total revenue for the past three years of firms to determine if reported firm size has changed or to compare the firm's total revenue with that from the DoD or other federal customers. Firms that supply goods and services in more than one industry make analysis more complex given that SBA size standards vary by industry and a business may be classified as large in one industry but small in another (Bounds, 2004).

We can analyze contract action data over time to determine if a supplier's status has changed. Yet until recently, actions on multiyear contracts originally won by a small business were recorded as small and counted toward meeting an agency's small-business goals for the duration of the contract, even if the business outgrew its size classification as a small business or was acquired by a large one (Zwahlen, 2007). One study cited estimated that, between FY 1998 and FY 2003, 30 percent of DoD small-business dollars ended up at large defense companies (Bounds, 2004). As of June 30, 2007, the government requires that contractors update their size status on long-term contracts before the fifth year and before exercising any option thereafter, following execution of a novation[8] agreement, or following a merger or acquisition of the contractor. DoD was unable to implement this change until FPDS was ". . . able to count dollars as awarded to a small business until the point when the concern represented that it was now no longer small, when dollars awarded after that point would be counted as dollars awarded to other than small business" (Assad, 2008). These modifications have since been implemented. As of March 25, 2008, DoD has mandated that changes to small-business size status be updated.

We used annual extracts of Dun & Bradstreet data on companies for FY 2001 to FY 2006, also used by the DoD during this time to validate contract action data, to help determine the parents of contractors and identify small businesses that were no longer in business or had been acquired or merged into another firm.

---

8  A *novation* is a successor in interest (Federal Acquisition Regulation SUBPART 242.1204—see General Services Administration, 2008).

We also analyzed DD350 and CCR data for contractors with at least one contract action coded as small business to see if they have

- remained under the threshold used to define a small business for all the NAICS codes for which they are registered in the CCR to provide goods and services
- graduated in selected NAICS codes in which they provide goods and services to the DoD but not others, which we classify as mixed
- graduated in all NAICS codes in which they provide goods and services to the DoD and are now not small
- been acquired or merged, gone out of business, or stopped seeking federal contracts.

We began our analysis of graduation with all contractors having at least one contract action coded as going to small business in the past 11 years. This yielded 165,170 unique small-business contractors between FY 1997 and FY 2007. Table 8.5 summarizes our findings from the CCR. Columns in the table provide

- CCR status: whether a contractor remains active or inactive in federal contracting
- number of contractors: the number of contractors by status
- percentage: the percentage of all contractors by status.

We first sought to locate the 165,170 contractors with small-business actions between FY 1997 and FY 2007 in the May 2008 CCR. Of these, we found that 72,505, or 43.9 percent of all contractors with a small-business action between FY 1997 and FY 2007, were not listed as active in the May 2008 CCR.[9] These contractors were either out of business, had been acquired, or were no longer pursuing federal contracts. Of the remaining 92,665, or 56.1 percent, contractors with a small-business action between FY 1997 and FY 2007 listed as active in the May 2008 CCR, 72,170 (43.7 percent) were listed as small for all their listed NAICS codes, indicating that they had not graduated. Another 14,266 contractors (8.6 percent) no longer had NAICS codes qualifying them as small. These contractors had either graduated, been acquired by a large firm and kept their original DUNS number, or originally been miscoded

**Table 8.5**
**Status of Contractors with Small-Business Actions in the May 2008 CCR Between FY 1997 and FY 2007**

| CCR Contractor and NAICS Status | No. of Contractors over 11 Years | Percentage of Contractors |
|---|---|---|
| Inactive | 72,505 | 43.9 |
| Active | 92,665 | 56.1 |
|     Small | 72,107 | 43.7 |
|     Not small | 14,266 | 8.6 |
|     Mixed | 6,292 | 3.8 |
| Total | 165,170 | 100.00 |

[9]   More precisely, 62,053 were not listed in the May 2008 CCR, and 10,452 were listed but not as "active" contractors.

as small. The remaining 6,292 contractors (3.8 percent) were listed in the May 2008 CCR as small in some NAICS codes but not others. These contractors may have grown, acquired other business, or become a part of a larger company that is small in some of its NAICS codes.

We next analyzed the D&B data for 2001 to 2006. We checked all 165,170 contractors that had a small-business contract action between FY 1997 and FY 2007 against their small-business status for their primary NAICS code as listed in the D&B data. We also checked whether they had the same or a different parent DUNS number than that listed in the CCR. Table 8.6 expands on Table 8.5 and summarizes this additional information, including

- contractor and NAICS status, including separate columns for status in the CCR and in the D&B data
- number of contractors: the number of contractors by CCR or D&B status
- percentage of contractors by status.

Nearly 92 percent of the 72,505 contractors not listed as "active" in the CCR had been active in the 2001–2006 D&B data. Of those that were active in the D&B data, 97 percent were listed as having the same parent as in the CCR data. This suggests that they had not changed parents in the past seven years. Three percent, or 1,907, of the remaining inactive CCR and active D&B contractors had one or more parents listed in the D&B data that differed from those in the CCR data. This suggests that these contractors may have participated

**Table 8.6**
**Evidence of Small-Business Graduation**

| Contractor and NAICS Status | | No. of Contractors[a] | Percentage of Contractors[a] | Possible Implication |
|---|---|---|---|---|
| 2008 CCR | 2001–2006 D&B | | | |
| Inactive | | 72,505 | 43.9 | Out of business/not seeking federal business |
| | Active | 66,479 | 91.7 | |
| | Same parent | 64,572 | 97.1 | |
| | 2 parents | 1,848 | 2.8 | Merger or acquisition |
| | 3–4 parents | 59 | 0.1 | Merger or acquisition |
| Active | | 92,665 | 56.1 | Seeking federal business |
| . Small | | 72,107 | 43.7 | Not graduated |
| Not small | | 14,266 | 8.6 | |
| | Active | 14,029 | 98.3 | |
| | Small | 3,485 | 24.8 | Recent graduation |
| | Mixed | 5,147 | 36.7 | Phased graduation |
| | Not small | 5,397 | 38.5 | Earlier graduation or error |
| Mixed | | 6,292 | 3.8 | |
| Total | | 165,170 | 100.00 | |

SOURCES: FY 1997–2004 DoD-wide DD350 data for transactions greater than $25,000; FY 2005–2007 FPDS-NG data for transactions greater than $2,500.

[a] Unique company I.D. codes.

in a merger or acquisition affecting their small-business status and precipitating a change in their DUNS number.

Virtually all, or 98 percent, of active contractors that were other-than-small in the May 2008 CCR data were active in the 2001–2006 D&B data. Twenty-five percent, or 3,485, of the active, other-than-small contractors were listed as small in the D&B data. This suggests that they may have graduated very recently from small-business status. Thirty-seven percent of the active, other-than-small CCR contractors were listed as both small and not small in the D&B data. Thus, they may have graduated several years earlier. Thirty-nine percent of the contractors that were active and not small in the CCR data were also active and not small in the D&B data. This indicates that they may have graduated before 2001 or the contract action(s) listing them as small may have been erroneous.

Additional data would be required to better understand mergers and acquisitions. The CCR contains only contractors that are registered to do business with the federal government. Many large companies are the parent of multiple and diverse smaller contractors. For example, Berkshire Hathaway Inc., a large, conglomerate with a primary NAICS of 524130, Reinsurance Carriers, and a parent of some DoD contractors, is not registered with the CCR. Although some of its subsidiaries are registered, the CCR contains no data on their parent firm, Berkshire Hathaway, other than its DUNS number. Of the 9,997 active contractors in the CCR with different parent DUNS numbers, we could not find information on the parent firms other than the DUNS number for 6,012 or 60 percent. The D&B data included information on all contractors and their parents, but we were told by the Director, Statistical Information Analysis Division, Defense Manpower Data Center that the DoD no longer purchases such data because of the transition to the FPDS-NG.

## Success of Small Businesses in Gaining Non-DoD Business

Most firms grow by expanding their base of customers. Indeed, some leading companies have programs to help their small and diverse suppliers grow.[10] We therefore analyzed how well contractors receiving small-business contract actions from the DoD have done in gaining other public and private revenue.

CCR data contain each contractor's average annual revenue for the previous three years. This information can be used to determine if a contractor has non-DoD revenue. For each of the 92,665 contractors with at least one small-business contract action in the past decade listed in the May 2008 CCR, we identified 85,044 that had contract actions in the FPDS between FY 2005 and FY 2007. We compared their total average annual revenue to their average annual revenue from DoD and federal contract actions for the past three fiscal years. We classified the contractors into two major groups: those that had the same parent and those that had a different parent as identified by our earlier analysis. We further classified each major group into those that were small in all NAICS codes, those that were small in some but not small in other

---

[10] For example, United Technologies Corporation shares scale-augmentation strategies with its diverse suppliers and Gillette helps some of its diverse suppliers expand globally (Varmazis, 2006). GlaxoSmithKline also helps its diverse suppliers in the United States find opportunities and partners in other countries (Teague, 2008). Part of Intel's strategy for expanding its global supplier diversity program is to work with them to expand their business (Atkinson, 2008).

NAICS industries, and those that were not small in all NAICS industries. Table 8.7 summarizes our findings. Table columns include information on

- contractor DoD dollars: comparison of the contractor's average annual total revenue to average annual DoD contract dollars, including
  - none: no DoD contract action dollars
  - less: average annual contract dollars less than average annual revenue
  - more: average annual revenue greater than average annual contract dollars
- contractor federal dollars: comparison of the contractor's average annual total revenue to average annual federal contract dollars, including
  - none: no federal contract-action dollars
  - less: average annual contract dollars less than average annual revenue
  - more: average annual revenue greater than average annual contract dollars
- total number of contractors: total number of contractors for which comparisons are available
- same parent: contractors whose CCR parent is the same as an earlier D&B parent
  - small: contractor is small in all industries
  - mixed: contractor is small in some industries but not others
  - not small: contractor is not small in any industry
- new parent: contractors whose CCR parent differs from an earlier D&B parent, that is, those that have been acquired by another firm
  - small: contractor is small in all industries

**Table 8.7**
**DoD Small Contractors' Non-DoD and Nonfederal Revenue**

| Total Contractor CCR Revenue Compared to: | | Total Contractors | Same Parent | | | New Parent | | |
|---|---|---|---|---|---|---|---|---|
| Contractor DoD Dollars | Contractor Federal Dollars | | Small | Mixed | Not Small | Small | Mixed | Not Small |
| None | None | 38 | 24 | 1 | 4 | 6 | 0 | 3 |
| None | Less[a] | 140 | 103 | 10 | 16 | 6 | 2 | 3 |
| None | More | 5,215 | 3,776 | 303 | 549 | 198 | 50 | 339 |
| Less[a] | Less[a] | 6,046 | 5,088 | 289 | 380 | 178 | 34 | 77 |
| Equal | Equal | 38 | 34 | 0 | 4 | 0 | 0 | 0 |
| More | Less[a] | 801 | 555 | 143 | 54 | 22 | 10 | 17 |
| More | More | 72,766 | 52,130 | 4,216 | 6,423 | 3,734 | 853 | 5,410 |
| % of DoD contractors with nonfederal dollars | | 85.57 | 84.48 | 84.97 | 86.45 | 90.11 | 89.88 | 92.49 |
| Total no. of contractors | | 85,044 | 61,710 | 4,962 | 7,430 | 4,144 | 949 | 5,849 |

SOURCES: FY 2005–2007 FPDS-NG and May 2008 CCR.

NOTE: The table includes DoD contractors coded as small business between FY 1997 and FY 2007 that were present in the FY 2005–2007 FPDS-NG and active in the May 2008 CCR (i.e., not expired).

[a] May reflect underreporting to CCR.

  – mixed: contractor is small in some industries but not others
  – not small: contractor is not small in any industry.

Of the 85,044 contractors active in the 2008 CCR and in the FPDS for FY 2005 to FY 2007, 38 had zero dollars for their DoD and other federal contract actions. Some contractors, 6,987, reported less annual revenue than their annual DoD or total federal revenue, suggesting that they may be underreporting their average annual revenue (or, possibly, losing money on nongovernment ventures). One group of 38 contractors reported average annual revenue that equaled their DoD and total federal revenue (i.e., they had no non-DoD federal revenue). Eighty-six percent, or 77,981, of the contractors we analyzed reported average annual revenue greater than their average annual DoD or federal contract action revenues. This suggests that they had nonfederal business before doing business with the federal government or have broadened their customer base.

For companies with the same parent over time, representing 86 percent of all contractors, we analyzed sources of revenue for small, mixed, and not small businesses, indicating their competitiveness in the overall economy. Table 8.8 summarizes our results, including

- statistics for sources of revenue: the mean and median percentages of DoD and nonfederal revenue income for contractors
- percentage of total DoD contractor revenue: the percentage of contractor revenue from DoD or a nonfederal source for contractors with consistent parents that were
  – small in all industries
  – mixed, or small in some industries but not others
  – not small in any industry.

The vast majority of DoD's recent contractors with the same parent over time obtain more than 80 percent of their revenue from nonfederal sources. Contractors that are not small had

**Table 8.8**
**Distribution of DoD Contractor Revenue by Source**

| Statistics for Sources of Revenue | Percentage of Total DoD Contractor Revenue: Same Parent | | |
|---|---|---|---|
| | Small | Mixed | Not Small |
| DoD | | | |
|   Mean | 10.80 | 14.47 | 7.03 |
|   Median | 1.48 | 2.02 | 0.42 |
| Nonfederal | | | |
|   Mean | 87.37 | 80.01 | 91.41 |
|   Median | 97.87 | 95.69 | 99.30 |

SOURCES: FY 2005–2007 FPDS-NG and May 2008 CCR.

NOTES: The table includes DoD contractors coded as small business between FY 1997 and FY 2007 that were present in the FY 2005–2007 FPDS-NG and active in the May 2008 CCR (i.e., not expired). Excluded are contractors with a negative percentage of non-DoD federal dollars and nonfederal dollars greater than 100 percent.

the largest shares of income from nonfederal sources, with most getting less than 1 percent of their revenue from federal sources. Similarly, most small businesses receive only about 2 percent of their revenues from federal sources. This suggests that DoD contractors have been successful at obtaining nonfederal customers, which should help support their eventual graduation from small-business programs.[11]

## Conclusions and Recommendations

The distribution of contracts and contract dollars among small-business contractors over the past 11 years varies considerably. A relatively small subset of contractors receives a large portion of dollars and contracts, with an even smaller portion receiving contract actions in every year we analyzed. Most small contractors receive few contracts for small amounts. Such infrequent and relatively small contract actions alone are not likely to help small businesses become large businesses.

Of the population of contractors that received one or more contract actions over the past seven years, 43 percent had not graduated. Forty-four percent of DoD small contractors over the past seven years were not active in the CCR in 2008. This suggests that they were out of business, had been acquired, or were no longer seeking federal contracts. Nine percent of contractors were no longer small, indicating that they had outgrown their size standards, had been acquired by a large business while keeping their original DUNS number, or had been incorrectly coded as small. The remaining 9 percent of small businesses we analyzed were mixed businesses, small in some industries but not in others.

Most DoD small contractors who have had the same parent for the past decade received a significant portion of their revenue, almost 90 percent, from non-DoD sources. This suggests that many of DoD's small contractors have been successful in broadening their customer base, which can facilitate growth and graduation.

To better understand small-business growth, graduation, and acquisition, the CCR needs to be annually archived so that small-business revenue, growth, mergers, and acquisitions can be tracked. Because the CCR lacks information on parent companies not seeking federal contracts, annual D&B company data would need to be purchased to analyze the complex relationships of some of its contractors. Finally, additional analysis is needed to understand the links between contracting practices and small-business graduation, particularly whether larger or longer contracts would be more likely to lead to graduation.

---

[11] It is also possible that such businesses rely on support from small-business preference programs of local and state governments or commercial firms beyond the scope of our analysis.

# CHAPTER NINE

# Conclusions and Implications

Our research suggests several areas for further attention in reducing impediments to small-business opportunities with the Department of Defense.

Analyses of Economic Census and DoD spending data show that the DoD spends more in industries where small businesses are less represented and less in industries where small businesses are more prevalent. Small businesses may be less prevalent in industries of greatest concern to the DoD because of industry consolidations (e.g., mergers and acquisitions) or industry dynamics such as large capital requirements to effectively compete. Such dynamics are beyond the ability of the DoD or its large contractors to control. Changes in industry classifications over time (e.g., from the Standard Industrial to the NAICS and across differing versions of the NAICS) also make it difficult to assess changing industry dynamics. As a consequence of underrepresentation of small businesses in industries where the DoD spends the most, the DoD sometimes has to use small businesses in a larger proportion than their industry representation to meet its overall small-business goals.

Policymakers may wish to reconsider what truly defines a small business. This might include revising size thresholds to recognize the large scale of production or investment required for production in some industries. In extreme cases, this might even include designating as "small" all firms not dominant within an industry and providing small-business preferences to those able to bring to market innovations sought through such policies. Alternatively, small-business goals could be set by industry rather than for all spending to account for the variance in small-business representation by industry. Implementing any remedies would require care, given how broad some industries are. For example, aircraft manufacturing includes large and sophisticated military aircraft as well as smaller general aviation craft. Detailed industry research beyond the indicators of the Economic Census would be required to set industry-specific thresholds or goals. Other industry features such as globalization or electronic commerce helping to reshape market dynamics should also be reviewed as new Economic Census and other data become available.

One evolving commercial practice requiring particular attention is the consolidation of requirements, contracts, and suppliers to reduce total costs, improve supply base performance, and enable supplier integration and supply chain synchronization. The federal government's Strategic Sourcing and the DoD's Performance-Based Logistics initiatives seek to adapt these practices to government purchases while avoiding consolidation that would exclude the participation of small business, what Congress defines as "bundling." Although there is great concern over the extent of bundling, the literature suggests that the extent of bundling, as measured by small-business protests regarding the practice filed with the SBA, is not widespread. Consolidating requirements over time into fewer, longer-term contracts with small businesses may increase the perception of bundling, because there are fewer bidding opportunities and, hence, fewer contracts for small businesses to win. Yet small businesses that win such contracts also realize the benefit of such consolidation through longer, more stable work. The DoD seeks to track the extent of contract "bundling" through data elements indicating whether the contract is the result of consolidating requirements. Unfortunately, the quality of these data is poor and does not provide a good picture of the current extent of DoD contract bundling. DoD may wish to explore ways to improve the quality of data on contract bundling.

Small businesses may have opportunities in subcontracting as well. Two areas of concern to Congress for small-business subcontracting are RDT&E and Professional Services. DoD spending in RDT&E has doubled in recent years and that for Professional Services has tripled. For RDT&E, our analyses of prime contract spending and the small-business share of that spending indicate that the value of contracts awarded to other-than-small businesses is increasing faster than that for small business. Although the proportion of contract dollars requiring subcontracting plans appears to be keeping pace with the growth in prime contract dollars, we cannot, unfortunately, determine whether the actual percentage of subcontracted dollars that are subject to a subcontracting plan is increasing. Detailed data on subcontracting that would help us determine the actual amount of RDT&E subcontracting dollars going to small businesses is not yet available for analysis.

The amount of prime contract dollars for Professional, Managerial, and Administrative Services is growing faster than the amount for RDT&E, but the DoD's total spend for these services in FY 2007 was $10 billion less than for RDT&E. Small businesses receive a much higher proportion of prime contract dollars for Professional Services than they receive for RDT&E, ranging from 30 percent or so for Managerial and Professional Services and up to 75 percent for Administrative Services. The percentage of contract dollars requiring a subcontracting plan in Professional Services has recently risen but, as noted, there are no currently available data for determining the amount of dollars that are actually subcontracted to small businesses. Once detailed data on subcontracting become electronically available, they should be analyzed for insights on subcontracting opportunities in these and other areas.

One way the DoD has sought to increase small-business participation in R&D is through the SBIR program, which funds R&D work in small businesses. Several studies over the years by the General Accounting Office, the NRC, and RAND have examined the DoD's SBIR program to determine its effectiveness along several dimensions, including how well it moves research results into larger DoD acquisition and technology development programs. Previous studies have identified three principal impediments to SBIR technology transition. First is the risk associated with technologies that are still immature as they emerge from the SBIR program. Acquisition program managers are reluctant to accept technologies still requiring significant time and money to integrate into larger systems or to place in the field. The second impediment is inadequate current resources for further developing immature technologies that emerge from the SBIR program. Third is that many acquisition program managers view the SBIR program as a "tax" and a distraction rather than a resource to be leveraged.

The federal government and the DoD have also used the Mentor-Protégé Program to introduce small businesses to opportunities. Although many protégés report that they benefit from the program and mentors are willing to participate as long as their costs are reimbursed, there is no quantitative evidence that the program directly helps boost the participation in DoD opportunities by businesses in the targeted categories. Perhaps future program funding and participation should be tied to evaluations linking program efforts to outcomes such as growth in revenues, number of customers, number of employees, and number of contracts as well as improvements in protégé quality, delivery, and cost performance.

Small businesses seeking opportunities with the DoD confront electronic payment systems that some may find confusing. Nevertheless, we found no clear evidence in the literature or in our interviews that DoD's Vendor Pay and other related e-commerce initiatives are barriers to most small businesses. The software and technology required for their use are relatively simple and inexpensive. Their use is also reported to improve the ability of small businesses to send the DoD invoices and receive payment.

We found no published quantitative analysis of small-business graduation. The literature on continuing consolidation in the defense aerospace industry and global trends toward collaborative, highly synchronized, lean supply chains reinforces an ongoing trend toward larger businesses. Increasing military specialization in technology such as stealth, which may have limited commercial applications and large capital costs, can also make it difficult for new entrants to succeed, let alone grow.

There are few incentives for DoD buyers and large contractors to help their small suppliers grow beyond small-business-size thresholds. Indeed, there are strong incentives not to do this, because doing so can result in the need to identify and qualify new small suppliers and make meeting small-business goals more difficult. Small businesses also have incentives to stay beneath the small-business-threshold size lest they lose their procurement preferences.

It is possible to assess small-business graduation over time through Individual Contract Action Reports and the Central Contractor Registration. Unfortunately, the CCR data on a firm's size by NAICS and average annual revenue are not annually archived, so determining whether a small contractor has graduated or has been acquired by a larger firm is very challenging. To better track the changing status of the government's small-business contractors, CCR data need to be archived annually and analyzed over time.

We were able to determine that 44 percent of small businesses receiving contract actions between FY 1997 and FY 2007 were no longer active in the May 2008 CCR. Of the 56 percent that were active, 43 percent were still small, 9 percent were not small, and 4 percent were classified as both small and not small depending on the industry in which they provided goods and services. Analysis of contracting practices, industries, and supply and service categories conducive to small-business growth could yield insights on how best to help more small businesses grow. CCR data do not provide identification information on the parent company of contractors that have been acquired unless the parent is also seeking business with the federal government. Thus, it is difficult to analyze the effects on contractors of mergers and acquisitions. Such an analysis would require additional data from Dun & Bradstreet on all parent companies of contractors.

Unfortunately, many of the data available for analysis are not ideal or no more than imperfect indicators of the impediments that interest Congress. Perhaps an even greater need for improving small-business opportunities than overcoming the impediments identified in this report is to devise data that can best analyze them. The effectiveness of small-business policies is not well understood and may require further research in particular on how policies shape the behavior of both beneficiaries and other, larger businesses (Gates and Leuschner, 2007).

# Data Availability and Quality Issues

In this appendix, we provide an overview of the data we used for the analyses in this document and discuss changes in key data elements, size and reporting thresholds, and the policies that underlie our analyses.

## Economic Census Data

The Census Bureau collects data on U.S. businesses every five years, obtaining information about establishments and firms, employment, labor costs, expenses, sales, assets, inventories, and capital expenditures. Our study uses the most recently available 2002 Economic Census data. We also use 1997 data. Data from the 2007 Economic Census data, the most recent, will not be available until late 2009 or 2010.

First, we discuss features of Economic Census data that must be modified for small-business analyses. Second, we note several business "rules" we developed for analysis of this data. Third, we note changes in Economic Census data that limit comparisons across time.

The SBA develops business-size thresholds that vary by industry for use in government programs, including which firms qualify as "small." For most industries, the SBA has set employment-based standard, usually 500, 750, or 1,500 employees per firm depending on the industry. For other industries, such as construction and transportation, it sets a sales-based threshold. These thresholds were used for our analyses of the Census data to calculate how prevalent small firms are within industries. Two industries have neither employment-based nor sales-based thresholds. Thresholds for utilities are based on megawatt hours, and those for financial institutions rely on an asset definition. Because we had no data on power generation or assets with which to gauge firm size in the Census data, these industries were excluded from our analysis.

We compared data on small-business prevalence by industry in the economy to DoD spending patterns as well as to non-DoD federal spending. For DoD spending, we use FY 2002 DD350 data that record the size of procurement awards and whether the contract is with a small firm.[1] It primarily applies to contracts greater than $25,000. These data feed into the Federal Procurement Data System database that contains contract transactions for almost all of the federal government.

Given the scope of our research, we were unable to correct data coding errors in the DD350 and FPDS data, including those for contract awards that contain purchases of both

---

[1] DD350 refers to the form number that is used to collect procurement award data for contracts.

goods and services but must be characterized with a single Product Service Code. Nevertheless, we do not believe that the data coding errors greatly affected our analyses.

Publicly available Economic Census data about firms are aggregated into cells, defined by industry code and size measured primarily using employment or sales. Table A.1 lists the cell categories for employment and sales.

A given cell, then, contains data on the number of firms and establishments within each employment or annual sales category for an industry. (Firms may have one or more establishments, referred to as contractors in the federal government, but are the controlling organizational entity. For example, someone owning a restaurant chain would own one firm and several establishments.)

The Economic Census data contain firm-level data for most industries. A few industries such as manufacturing, construction, and mining have only establishment-level data. We used firm data whenever possible to approximate SBA business-size determinations.

Given the size of the manufacturing sector and the amount of money the DoD spends in it, having only establishment-level data could pose challenges for our analysis. Yet Reardon and Moore (2005) found that manufacturing is dominated by large establishments and that their findings for manufacturing would not qualitatively change if they had analyzed firm data in manufacturing rather than establishment data, except that small businesses would be even less prevalent than they reported. They also found that industries that manufacture parts rather than larger goods (e.g., aircraft engines as opposed to aircraft) have a somewhat higher representation of small establishments.

Reardon and Moore (2005) also found that in construction, the number of firms and establishments is quite similar and concluded that having only establishment data for construction does not pose a problem. Further, although the differences in mining were greater, that industry is not a recipient of a great deal of DoD spending and thus using establishment rather than firm data does not materially change the overall conclusions.

Table A.1
Economic Census Employment and
Annual Sales Categories

| Employment | Annual Sales |
| --- | --- |
| 1–4 | <$100,000 |
| 5–9 | $100 K–$249 K |
| 10–19 | $250 K–$499 K |
| 20–49 | $500 K–$999 K |
| 50–99 | $1 M–$2.49 M |
| 100–249 | $5 M–$9.9 M |
| 250–499 | $10 M–$24.9 M |
| 500–999 | $25 M–$49.9 M |
| 1,000–2,499 | $50 M–$99.9 M |
| 2,500 or more | $100 M or more |

For confidentiality reasons, Economic Census data are not provided when only one or a few firms or establishments are in a cell or cells such that particular firms could be identified. Data for at least two cells are suppressed and indicated with NA for Not Available, but industry totals are still provided. Thus, we can calculate the total employment or annual sales for the suppressed cells. Reardon and Moore (2005) tested a number of different methodologies for estimating the missing data and came closest to replicating actual data when the suppressed data were imputed by calculating averages within size categories across industries defined by two-digit NAICS codes.

Economic Census employment and annual sales categories do not always align with SBA small-business threshold sizes. For example, in some industries the SBA threshold is 750 employees, but the Economic Census data report the characteristics for firms with 500 to 999 employees. Reardon and Moore (2005) found that assuming that firms were uniformly distributed within the category was a reasonable mid-range strategy for estimating the number of small businesses within an Economic Census cell. That is, the fraction of businesses that could be considered "small" was the same fraction as the SBA threshold relative to the Economic Census cell endpoints. Thus, a threshold that was halfway between the endpoints meant that half of the category's employment and sales was attributed to small firms.

We used Reardon and Moore's (2005) methods for imputing suppressed Economic Census cell data and for estimating the small-business portion of cell firms when the SBA threshold did not align with Economic Census cell size.

## Statistics of U.S. Businesses

We also used data from the Statistics of U.S. Businesses (SUSB), an annual database constructed by the U.S. Bureau of the Census. These data include establishment "demographics," that is, data on establishment births, deaths, expansions, and contractions, that are available for four-digit NAICS codes from 1998 to 2004 on the Web site of the U.S. Census Bureau.[2] The database refers to establishments, i.e., separate physical locations of firms, and covers all U.S. business establishments with employees, except for railroads and households. Information is provided for six 12-month periods each measured from March of the beginning year to March of the ending year:[3] 1998–1999, . . ., 2003–2004).[4] SUSB provide aggregated data for business establishments classified by firm size, where firm size in each year is measured by the aggregate nationwide employment of the parent firm to which each establishment belongs. These data document for each four-digit NAICS code for each of the six 12-month periods the following information, for six firm-size categories (1–4 employees, 5–9, 10–19, 20–99, 100–499, and 500+ employees).[5]

- the total number of establishments at the beginning of the "year," i.e., in March of the first year of the pair

---

[2]  U.S. Census Bureau (2008).

[3]  Specifically, the data for each year refer to the week that includes March 12.

[4]  Data for 1989–1998 are available for SIC codes.

[5]  Firm size is measured at the beginning of the "year" or at the time of the birth of the establishment.

- the number of births of establishments during the year, i.e., new establishments that began during the year; these can be new businesses or additional sites for a multisite business (such as Wal-Mart)[6]
- the number of deaths of establishments during the year, i.e., establishments that stopped operating during the year; a "death" could be due to a firm going out of business or to a business closing a site due to outsourcing, restructuring, merger, or acquisition
- the number of continuing establishments that expanded in size (i.e., had an increase in the number of employees) during the year
- the number of continuing establishments that contracted in size (i.e., experienced a reduction in the number of employees) during the year.

Information is also available on the total number of employees for each of these categories.[7]

## DD350 and FPDS-NG data

The Acquisition Advisory Panel (AAP) (2007) reported that, since its inception, the Federal Procurement Data System has been plagued with claims that its data are inaccurate. The GAO found that data problems were largely due to inaccurate or untimely data input by reporting agencies. The AAP further noted

> In general, it seems that FPDS-NG data at the highest level provides significant insight. However, the reliability of that data, especially on . . . new reporting elements, begins to degrade at the more granular level due to data specificity on elements for which those reporting may have less familiarity and training (Acquisition Advisory Panel, 2007, p. 439).

For some of our analyses, we use data collected using DoD Form 350, the Individual Contract Action Report (ICAR), from 1983 to 2004, for transactions of at least $25,000. Data collectors sent these data to the Statistical Information Analysis Division (SIAD) located within the Defense Manpower Data Center which consolidated, validated, and stored them in the DD350 Contract Action Reporting System and reported them annually to the FPDS. The data that were used for these analyses and their associated documentation are available in electronic form at the SIAD/DMDC Web site.[8]

For FY 2005 to FY 2007, we use data from the Federal Procurement Data System—Next Generation. Federal agencies, including the services and defense centers within the DoD, directly report their individual contract actions to a contractor-run General Services Administration Web site for the FPDS-NG. Starting in FY 2007, this was done through a machine-to-

---

[6]  Births are defined as establishments with either no record or no employment in the beginning of the period and a record with positive employment for the end of the period. Deaths are similarly defined as establishments with a record with positive employment for the beginning of the period and either no record or no employment at the end of the period (Armington, 1998).

[7]  For more information about the data, see Armington (1998).

[8]  Department of Defense (n.d.).

machine interface. These data and associated documentation are available in electronic form at the FPDS Next Generation Web site.[9]

Below, we discuss three issues regarding analysis of DD350 and FPDS-NG data. First, we note changes in data over time that limit comparisons across time. Second, we discuss features of the data that must be modified for analyses. Third, we note several business "rules" that we developed for analyzing these data.

## Changes in Individual Contract Action Reports

Before 1983, contract actions of $10,000 or more were required to be reported as DD350 actions. The reporting threshold was raised to $25,000 in 1983 (Statistical Information analysis Division, 2008). In FY 2005, FAR 4.602(c)(1) required that contract data collection points in each government agency collect data on all transactions over $2,500 (this was raised to $3,000 in FY 2007), meaning that the DoD now requires reporting of contract transactions exceeding $3,000 on DD350 forms. The Army began implementing the reporting of contract actions valued at more than $2,500 on DD350 forms before FY 2005 more aggressively than the other services and DoD agencies, but all military branches and agencies have reported at least some actions of less than $25,000 on DD350 forms since at least FY 2000. Starting in FY 2007, the services and defense centers directly report their DD350 individual contract actions data to FPDS-NG through a machine-to-machine interface (Lee, 2005).

### Phasing Out of DD1057

From 1964 until FY 2004, transactions below the DD350 reporting threshold were reported by each contracting organization on DoD Form 1057, in a monthly summary. It contained the total number of contract actions and total dollars broken out by various small-business categories. Before 1996, when separate reporting began, purchases made via the government purchase card were included in DD1057 reports. In FY 2005, DD1057 was no longer collected and all actions and dollars were reported by Form DD350, which feeds FPDS-NG (Statistical Information Analysis Division, 2008). Some organizations, including the Army, started phasing out DD1057 earlier and began reporting actions less than $25,000 on form DD350.

Because FPDS-NG included most, but not all, actions previously reported by DD1057, any analyses of DoD spending over time that included the shift to FPDS-NG need to adjust for the different contract reporting thresholds between DD350 and FPDS-NG.

To assess changes in the value of small contract actions by budget category over time, we assumed that DD1057 dollars were allocated by Federal Supply Class and Product and Service Code in years before FY 2005 as were dollars from contract actions for no more than $25,000 from FY 2005 to FY 2007.

### Governmentwide Commercial Purchase Card

The governmentwide commercial purchase card began as a response to Executive Order 12351 to "establish programs to reduce administrative costs and other burdens which the procurement function imposes on the Federal government and the private sector" (Executive Order 12352, 1982).

---

[9]  Federal Procurement Data System (n.d.).

It was introduced to simplify and speed the federal government's paper-based, time-consuming, and slow purchase order process for small dollar procurements (purchases of up to $25,000). Card utilization increased during the 1990s as a result of the "Reinventing Government" federal procurement initiative: Lower Costs and Reduce Bureaucracy in Small Purchases Through the Use of Purchase Cards (General Accounting Office, 1994b). The purchase card has become the government's primary payment and procurement method for purchases under $2,500 (raised to $3,000 in FY 2007 to adjust for inflation), which are often referred to as micro-purchases (Federal Register, 2006). Since 1984, the General Services Administration has been responsible for contracting for government charge card services (Mead, 2002).

### SIC to NAICS

The North American Industry Classification System has replaced the U.S. Standard Industrial Classification (SIC) system as the industry code in both the Economic Census and DD350/FPDS. The NAICS was developed jointly by the United States, Canada, and Mexico to provide new comparability in statistics about business activity across North America.

Industry codes changed in the ICAR with the replacement of the SIC in FY 2001 with the NAICS in 1997, making it difficult to compare codes for industries over time. NAICS codes were revised in 2002 with the new codes first being used in the ICAR in FY 2003. Another NAICS revision was done in 2007 but had not been implemented in the FY 2007 FPDS. Both the change from SIC to NAICS codes as well as revisions in NAICS codes limit comparisons of DD350 data by industry over time. They also limit comparisons of Economic Census data by industry over time. Indeed, the Census has not done any industry comparisons between 1992 and 1997 because of the change. It is only now starting to do some industry comparisons, but further changes make such comparisons problematic.

## Using Contract Action Data for Analyses

The contract action data were not originally designed for analyses. Rather, they were designed to assess compliance with several contracting regulations, including those for open competition as well as for procurement from small or small and disadvantaged businesses. As such, they must be supplemented with other data when used for spend analyses. We note below two particular issues we had to address in DD350 and FPDS-NG data: identifying parent companies of local contractors and challenges posed by missing data.

### Identifying Parent Companies of Local Contractors

The ICAR data identify individual contractors using a contractor identification code/number (i.e., DUNS number[10]), company name, and address. Aggregating ICAR data by individual contractor DUNS numbers alone would fail to reveal true spend totals with large corporations that have multiple locations or divisions, each with a unique contractor/DUNS number. Contract action data in the DD350 that we used for our analyses also contain the DUNS number for the ultimate parent of each contractor, i.e., the top level of the corporation, but lack any

---

[10] The Dun & Bradstreet DUNS number is a unique nine-digit identification code used to reference single business entities while linking corporate family groups together. It is an internationally recognized common company identifier in electronic data interchange and global electronic commerce transactions.

further information about that parent. The FPDS-NG data have a data field for parent DUNS number, but it was blank for the FPDS data we downloaded. Analysts must use outside sources to obtain the name of the parent company as well as characteristics such as its size. Although the FPDS-NG has a data element for a contractor's parent DUNS number, our analyses found that the data field is currently blank.

From FY 2001 to FY 2006, SIAD/DMDC supplied RAND with a Dun & Bradstreet database corresponding in time to the public release of the DD350 data on its Web page for those years, enabling us to identify the names and some characteristics of parent companies. Individual company Web pages also helped us to verify some linkages between contractors and major divisions of their parents. Coding errors in the local contractor and ultimate parent company variables appear to be minimal in more recent DD350 data, allowing comparisons over time. One reason for these minimal errors is post-year data scrubbing performed by SIAD/DMDC. With a machine-to-machine interface for reporting ICAR data to FPDS-NG, periodic verification of ultimate parent codes of local contractors may be needed to ensure consistency of data over time.

### Challenges Posed by Missing Data

Contract actions can include more than one Product Service Code. The ICAR format permits only one PSC to be coded. Standard practice is to record the PSC accounting for the most dollars for each action, which can lead to overstatement of dollars for some PSCs and understatement for others. The extent of error in dollar totals for this practice is unknown.

Some records are coded for PSC 9999, miscellaneous items. To better understand the purchases in this PSC the contractor's NAICS code for the action can be analyzed. NAICS codes can be less precise than PSCs regarding purchases but, for analytical purposes, they are superior to PSC 9999. Similarly, some ICARs were coded with the NAICS code for soybean farming (NAICS 111110), the first NAICS code on the master list of NAICS.[11] To better understand the contractor industry PSCs can be analyzed.

## Business Rules for DD350 Data

In analyzing DD350 data, we developed several rules to help ensure that our analyses would be replicable. We describe the primary ones below.

### Dollars

Dollars were summed across contract actions. This included negative amounts for deobligations (cancellation of an obligation to pay). In some cases, where indicated, we deleted deobligations from our analysis.

### Contracts

Contracts were counted by the number of distinct 15-position, alphanumeric character contract numbers across contract actions.

---

[11] The data erroneously indicated that the DoD spent $185 million in soybean farming between FY 2000 and FY 2005 (Acquisition Advisory Panel, 2007).

### Parent Companies

Using the parent company DUNS number in the DD350 data ensured that our dollar results exactly matched the tabulations available on the SIAD/DMDC procurement Web site.

### Small-Business Status

To identify the population of prospective small businesses supplying DoD with goods and services, we used the small-business codes as recorded on each ICAR. We removed those ICARs that did not count toward meeting the DoD's small business goals such as foreign military sales, overseas purchases, purchases of specified goods from UNICOR, and contracts awarded for the Defense Commissary Agency, among others. The FY 2007 FPDS-NG data that we downloaded in January 2008 for our analyses are not likely to match more recent versions of such data because of ongoing data cleansing and validation efforts. We were able to match the SIAD/DMDC tabulations for DoD small-business use. However, we observed inconsistent coding in the ICAR data for some contractors. This could be due to inaccurate coding or to a policy in place during our period of analysis that allowed contracting offices to continue to record small businesses that had either grown large or had been acquired by a large company as small for the life of a multiyear contract.

We used the Central Contractor Registration to validate the 2008 size status of those contractors from our original population of small businesses that were still registered to do business with the federal government.

### CCR Data

The CCR collects, validates, stores, and disseminates data in support of U.S. federal government agency acquisition missions. Registration in the CCR is required before the federal government will award contracts. Registrants are required to provide basic information relevant to procurement and financial transactions and to update or renew their registration at least once per year to maintain an active status. CCR validates each registrant's information, secures and encrypts the data, and electronically shares it with federal agencies' finance offices to facilitate paperless payments through electronic funds transfer and with federal government procurement and electronic business systems (Central Contract Registration, 2008b).

The 2008 CCR data we obtained are not as comprehensive as the D&B data SIAD/DMDC used to validate DD350 data. The CCR contains only contractors that are registered to do business with the federal government, and many large parent companies of DoD contractors do not themselves do business with the federal government and thus are not registered. For example, Berkshire Hathaway Inc., the parent of several diverse DoD contractors with their own CCR listings, is not listed in the CCR. Thus, other than Berkshire Hathaway's DUNS number linked to the contractors, there is no additional information on them in the CCR such as their primary NAICS (524130, Reinsurance Carriers).

Consequently, of the 9,997 contractors with different parent DUNS numbers we could not find any information on their parents in the CCR other than a DUNS number for 6,012 (60 percent) of them.

Information on all contractor parents (e.g., name and size) was included in the D&B data that SIAD/DMDC used to use. Not so with the CCR we were able to access.

The SBA uses data on firm size to certify an entity as a small business.

**D&B Data**

Dun & Bradstreet's global commercial database contains more than 125 million companies—77 million active and available for risk, supply, sales and marketing, and e-business decisions and 38 million inactive available with historical information for file-matching and data-cleansing purposes. D&B data link the D&B DUNS numbers of parents, subsidiaries, head-quarters, and branches for more than 70 million corporate family members around the world. Of these company records, 8,745,967 company records have a family tree. Further, 80 percent of their active U.S. file contains businesses with ten or fewer employees.[12]

SIAD/DMDC had ordered special extracts of these data that included DoD contractors and their parents.

**Rules for Counting Small Businesses on Multiyear Contracts**

Until June 29, 2007, contracting organizations were allowed to count as small contract actions on multiyear contracts that were originally awarded to a small contractor even if the size of that contractor had changes during the duration of the contract because of business growth or merger with or acquisition by another firm. Consequently, we did not attempt to validate the size of contractors on actions that were coded as going to a small business.

## Summary of Changes in Data Elements and Policies That Make Analyses over Time Very Challenging

As we discuss above, several key data elements in the Economic Census and DD350/FPDS-NG have had their coding structure completely revised in recent years, which makes comparisons over time particularly challenging. Additional changes in size and reporting thresholds and reporting policies further complicate analyses over time. We summarize many of these changes below.

- The threshold for ICAR and DD1057 data collection increased from $10,000 to $25,000 in FY 1983. It dropped to $2,500 in FY 2005 and the DD1057 data were no longer collected. The threshold was raised to $3,000 in FY 2007 for FPDS-NG.
- Standard Industry Codes used in the 1992 Economic Census changed to North American Industry Classification System (NAICS 1997) for the 1997 Economic Census. Then NAICS 1997 changed to NAICS 2002, which were used for the 2002 Economic Census and NAICS 2002 were changed to NAICS 2007 for the 2007 Economic Census. Further, ICAR data changed from SIC to NAICS 1997 in FY 2001 and changed again from NAICS 1997 to NAICS 2002 in FY 2003.
- DD1057 was phased out in 2005 and DD350 shifted to FPDS-NG with different formats, different reporting thresholds, and different data elements.
- ICARs below $25,000 were phased into DD350 data by some organizations such as the Army and Marine Corps before FY 2005.
- The ICAR threshold for FPDS-NG was >$2,500 in FY 2005 and shifted to ICARs >$3,000 in 2007.

---

[12] For more on D&B registrants, see Dun and Bradstreet, home page, n.d.

- Rules for calculating goaling dollars have changed over the years.
- Rules for counting small-business dollars on multiyear contracts have changed.
- Small-business size thresholds vary by NAICS and over time.
- FPDS parent DUNS numbers are not currently populated.
- CCR data are not archived and do not include information beyond a DUNS number for parents that are not registered to do business with the federal government.
- Entities may be referenced by differing terms, not all of which may be familiar to DoD personnel or others concerned with these issues. Figure A.1 summarizes some varying terms for entities of interest to small-business analyses.

**Figure A.1**
**Terms Used for Enterprise Relationships**

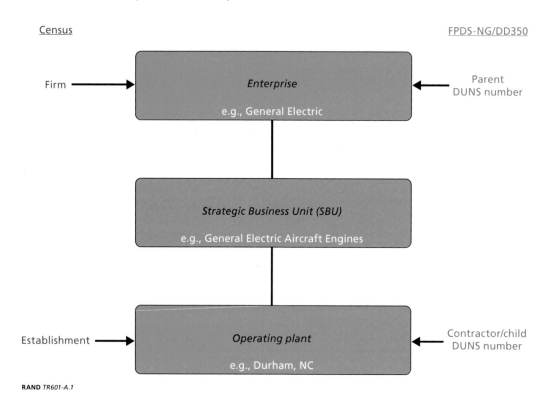

RAND TR601-A.1

# References

Acquisition Advisory Panel, Final Report, January 2007. As of June 23, 2008:
http://www.acquisition.gov/comp/aap/24102_GSA.pdf

Affourtit, Barb, email to Nancy Y. Moore, RAND, regarding Datafile Definitions, December 17, 2003.

Aitoro, Jill R., "Bundled Deals: A Lose-Lose for Small Biz," VARbusiness, April 17, 2006.

Al-Qirim, N.A.Y., and B. J. Corbitt, "Determinants of Electronic Commerce Usage in Small Businesses in New Zealand," 12th European Conference on Information Systems, Turku, Finland, June 14–16, 2004.

Armington, Catherine, "Statistics of U.S. Businesses—Microdata and Tables," June 4, 1998. As of June 23, 2008:
http://www.presidency.ucsb.edu/ws/print.php?pid=42283

Assad, Shay D., "Small Business Size Rerepresentation—FAR Case 2006-032," March 25, 2008. As of June 23, 2008:
http://www.acq.osd.mil/dpap/policy/policyvault/2008-0228-DPAP.pdf

Atkinson, William, "Intel Expands Supplier Diversity Program Globally," *Purchasing,* January 17, 2008. As of April 23, 2008:
http://www.purchasing.com/article/CA6518800.html

"At Today's Cisco Systems, the Fewer Suppliers the Better," *Purchasing,* April 20, 2006. As of April 23, 2008:
http://www.purchasing.com/article/CA6324660.html

Auguste, Byron G., Eric P. Harmon, and Vivek Pandit, "The Right Service Strategies for Product Companies," *The McKinsey Quarterly,* February 2006.

Bail, Philip G., Jr., "Socioeconomic Programs—On the Road to Failure?" *Contract Management,* Vol. 46, No. 4, April 2006, pp. 46–51.

Baker, Robert-Allen, *Mining the Small Business Resource—Issues and Recommendations; A White Paper,* Washington, D.C.: The Small Business Technology Council, January 2007. As of May 21, 2008:
http://www.nsba.biz/docs/sbir_white_paper_iv_final_11_jan_07.pdf

Baldwin, Laura H., Frank A. Camm, and Nancy Y. Moore, *Federal Contract Bundling: A Framework for Making and Justifying Decisions for Purchased Services,* Santa Monica, Calif.: RAND Corporation MR-1224-AF, 2001. As of April 24, 2008:
http://www.rand.org/pubs/monograph_reports/MR1224/

Baum, J. Robert, "Entrepreneurs' Start-up Cognitions and Behaviors: Dreams, Surprises, Shortages and Fast Zigzags," May 2004. As of May 21, 2008:
http://www.babson.edu/entrep/fer/BABSON2003/XXV/XXV-P1/xxv-p1.htm

Bowman, Robert, "For One Aerospace Supplier, Small Is Anything But Simple," Global Logistics & Supply Chain Strategies, March 1, 2007. As of May 25, 2008:
http://www.supplychainbrain.com/content/headline-news/single-article/article/for-one-aerospace-supplier-small-is-anything-but-simple/

Bounds, Gwendolyn, "Why U.S. Contracts for Small Businesses Go to Big Companies," *Wall Street Journal,* November 9, 2004.

Brewin, Bob, "Defense Says Bye Bye to EDI," *The Government Executive,* August 15, 2007.

Business Transformation Agency, "EDA User Guide," Department of Defense, Washington, D.C., 2008a. As of May 22, 2008:
http://eda.ogden.disa.mil/eda_main.htm

———, "Setting Up Your Machine for WAWF," Department of Defense, 2008b. As of May 22, 2008:
https://wawf.eb.mil/

Byrne, John A., "The Corporation of the Future," *Business Week,* August, 31, 1998.

Cabreira-Johnson, Debbie, Program Director, Procurement Technical Assistance Center, Los Angeles County Office of Small Business, email to RAND, "RE: DoD Small Business Study," March 18, 2008.

Central Contractor Registration, "Central Contractor Registration," 2008a. As of May 21, 2008:
http://www.ccr.gov/

———, "Central Contractor Registration Handbook," May 2008b. As of June 23, 2008:
http://www.ccr.gov/doc/CCR_Handbook.pdf

Cho, Vincent, "Factors in the Adoption of Third-Party B2b Portals in the Textile Industry," *The Journal of Computer Information Systems,* Vol. 46, No. 3, Spring 2006, pp. 18–31.

Clark, Major III, and Chad Moutray, "The Future of Small Businesses in the U.S. Federal Government Marketplace," *Journal of Public Procurement,* Vol. 4, No. 3, 2004, pp. 450–470.

Clark, Major III, Chad Moutray, and Radwan Saade, "The Government's Role in Aiding Small Business Federal Subcontracting Programs in the United States," September 2006. As of May 5, 2008:
http://sba.gov/advo/research/rs281tot.pdf

Code of Federal Regulations, January 1, 2003. As of September 15, 2008:
http://www.cfsan.fda.gov/~lrd/9cf362.html

———, Title 13, Vol. 1, revised as of January 1, 2008. As of May 3, 2008:
http://www.access.gpo.gov/nara/cfr/waisidx_08/13cfrv1_08.html

Cooper, Robert G., and Scott J. Edgett, "Stage Gate," Produce Development Institute, Inc., 2008. As of June 4, 2008:
http://www.prod-dev.com/stage-gate.shtml

Coulter, Dawn, Director, eSolutions, Defense Finance and Accounting Service, email to RAND with attachments, "RE: Study on Impact of Vendor Pay on Small Business" (trouble ticket data provided by the WAWF Program Manager), March 7, 2008.

Crain, W. Mark, and Thomas D. Hopkins, "The Impact of Regulatory Costs on Small Firms," Small Business Administration Office of Advocacy, September 2005. As of May 23, 2008:
http://www.sba.gov/advo/research/rs207.pdf

DeBeer, Marthin, "Innovation Through Internal Venturing," *ETL Quarterly,* Vol. 3, No. 4, September 26, 2006, pp. 5–6.

Defense Acquisition University, Defense Acquisition Guidebook, Table 10.5.2.1, TRL Descriptions, Ft. Belvoir, Va.: Defense Acquisition University Press, December 2004. As of June 5, 2008:
https://akss.dau.mil/dag/DoD5000.asp?view=document&rf=Guidebook/IG_c4.3.2.4.3.asp

———, "Integrated Defense Acquisition, Technology and Logistics Life Cycle Management Framework," 2008, p. 8. As of May 26, 2008:
https://acc.dau.mil/IFC/Back_pg8.htm

Defense Finance and Accounting Service, DFAS Columbus Center, *Contract Pay Information Handbook,* Columbus, Ohio, January 2006a. As of April 30, 2008:
http://www.dfas.mil/contractorpay/ContractPayInformation.pdf

———, *Electronic Data Interchange Guide,* Columbus, Ohio, November 3, 2006b.

Defense Finance and Accounting Service, eSolutions Office, *Wide Area Workflow Getting Started Guide,* version 3.0.12, August 14, 2007. As of May 21, 2008:
http://www.dfas.mil/contractorpay/electroniccommerce/ECToolBox/WAWFVendorGettingStartedGuide.pdf

Defense Finance and Accounting Service, "Contract/Vendor Pay," 2008. As of May 21, 2008:
http://www.dfas.mil/contractorpay.html

Defense Logistics Agency, *DLIS Web Enabled Products—CAGE Codes,* Defense Logistics Information Service, 2008. As of May 22, 2008:
http://www.dlis.dla.mil/cage_welcome.asp

Dennis, William F., Jr., "NFIB National Small Business Poll: Success, Satisfaction, and Growth," 2001. As of June 5, 2008:
http://www.411sbfacts.com/files/success.pdf

Department of Defense, DoD Procurement: Procurement Reports and Data Files for Download, n.d. As of September 15, 2008:
http://siadapp.dmdc.osd.mil/procurement/Procurement.html

———, DoD Financial Management Regulation 7000.14-R, Volume 2B, Budget Formulation and Presentation (Chapters 4–19), June 2006. As of April 16, 2008:
http://www.defenselink.mil/comptroller/fmr

Department of Defense Office of Small Business Programs, *Mentor-Protégé Program Annual Reports to Congress,* annual.

———,"Mentor-Protégé Program Participant Highlights," August 14, 2007. As of April 25, 2008:
http://www.acq.osd.mil/osbp/mentor_protege/nunnperry/sstories.htm

———, "Program Goals and Statistics," March 17, 2008. As of May 5, 2008:
http://www.acq.osd.mil/osbp/statistics/goals.htm

De Rugy, Veronique, "Let Them Be," *Wall Street Journal,* April 23, 2007.

"Differences of Opinion Fail to Taint Benefits of Outsourcing," *World Airline News,* July 21, 2000.

Dobbs, Matthew, and R. T. Hamilton, "Small Business Growth: Recent Evidence and New Directions," *International Journal of Entrepreneurial Behaviour and Research,* Vol. 13, No. 5, 2007, pp. 296–322.

Duffy, Roberta J., "Being Diverse, Within and Throughout the Chain," *Inside Supply Management,* Vol. 15, No. 7, July 2004. As of May 15, 2008:
http://www.ism.ws/files/SR/July04SRrprnt%2Epdf

Dun & Bradstreet, *About D&B,* home page, n.d. As of September 15, 2008:
http://www.dnb.com/US/about/index.html?cm_re=HomepageB*AboutDB*AboutDBLink

———, *D&B Today,* home page, n.d. As of September 15, 2008:
http://www.dnb.com/us/about/company_story/dnbdbtoday.html

———, *Facts & Figures,* home page, n.d. As of September 15, 2008:
http://www.dnb.com/us/about/db_database/dnbstatistics.html

———, "DUNS Request Service," 2008. As of May 21, 2008:
http://fedgov.dnb.com/webform

Dunn, Jim, "Small Partners, Big Business," *Government VAR,* March 8, 2004. As of March 20, 2008:
http://www.crn.com/government/18842075

Eagle Eye Publishers, Inc., "The Impact of Contract Bundling on Small Business, FY 1992–FY 2001," October 2002. As of April 24, 2008:
http://www.sba.gov/advo/research/rs221tot.pdf

El Tarabishy, Ayman, "NFIB National Small Business Poll: IT Issues," 2007. As of May 21, 2008:
http://www.411sbfacts.com/files/V7I5_poll_ITfinal%20(2).pdf

Elia, E., L. A. Lefebvre, and É. Lefebvre, "Typology of b-to-b e-Commerce Initiatives and Related Benefits in Manufacturing SMEs," Big Island, Hawaii: 37th Annual Hawaii International Conference on System Sciences, January 5–8, 2004, pp. 70167b–70175b.

ESIS, Honeywell Case Study, 2005. As of May 22, 2008:
http://www.esisinc.com/aboutus/casestudy_honeywell.php

———, Cessna Support, Cessna letters to suppliers, 2008. As of May 22, 2008:
http://www.esisinc.com/support/cessna/index.php

Executive Order 12352, Federal Procurement Reforms, March 17, 1982. As of June 23, 2008:
http://www.presidency.ucsb.edu/ws/print.php?pid=42283

Federal Procurement Data Center, Next Generation Web site, n.d. As of September 15, 2008:
https://www.fpds.gov

Federal Register, "Federal Acquisition Regulations: Final Rules," September 28, 2006. As of June 23, 2008:
http://edocket.access.gpo.gov/2006/06-8199.htm

Fields, Gary, "Security Vetting of Employees Is Highly Prized," *Wall Street Journal*, February 24, 2004.

Fisher, Julie, and Annemieke Craig, "Developing Business Community Portals for SMEs? Issues of Design, Development and Sustainability," *Electronic Markets*, Vol. 15, No. 2, May 2005, pp. 136–145.

Foreman, Tim J., e-mail to Nancy Y. Moore, RAND, April 9, 2008.

Garcia, Marie L., and Olin H. Bray, Fundamentals of Technology Roadmapping, Albuquerque, N.M.: Sandia National Laboratories, SAND97-0665, April 1997. As of April 15, 2008:
http://www.sandia.gov/PHMCOE/pdf/Sandia'sFundamentalsofTech.pdf

Gates, Susan M., and Kristin J. Leuschner, eds., *In the Name of Entrepreneurship? The Logic and Effects of Special Regulatory Treatment for Small Businesses*, Santa Monica, Calif.: RAND Corporation, MG-663-EMKF, 2007. As of June 5, 2008:
http://www.rand.org/pubs/monographs/MG663/

General Accounting Office, *Federal Research: Small Business Innovation Research Participants Give Program High Marks*, July 1987. As of May 21, 2008:
http://archive.gao.gov/d28t5/133554.pdf

———, *Assessment of Small Business Innovation Research Programs*, January 1989. As of May 15, 2008:
http://archive.gao.gov/d15t6/137756.pdf

———, *Small Business Innovation Research Shows Success But Can Be Strengthened*, March 1992. As of May 15, 2008:
http://archive.gao.gov/t2pbat6/146114.pdf

———, *Defense Contracting: Implementation of the Pilot Mentor-Protégé Program*, February 1994a. As of April 28, 2008:
http://archive.gao.gov/t2pbat4/150805.pdf

———, *Implementation of the National Performance Review's Recommendations*, December 1994b. As of June 23, 2008:
http://www.gao.gov/archive/1995/oc95001.pdf

———, *DoD's Small Business Innovation Research Program*, April 1997. As of May 21, 2008:
http://www.gao.gov/archive/1997/rc97122.pdf

———, *Defense Industry Consolidation and Options for Preserving Competition*, April 1998a. As of May 19, 2008:
http://www.gao.gov/archive/1998/ns98141.pdf

———, *Observations on the Small Business Innovation Research Program*, April 1998b. As of May 21, 2008:
http://www.gao.gov/archive/1998/rc98132.pdf

———, *Defense Contracting: Sufficient, Reliable Information on DoD's Mentor-Protégé Program Is Unavailable*, March 1998c. As of April 28, 2008:
http://www.gao.gov/archive/1998/ns98092.pdf

———, *Best Practices: Better Management of Technology Development Can Improve Weapon System Outcomes*, July 1999. As of May 21, 2008:
http://www.gao.gov/archive/1999/ns991620.pdf

————, *Contract Management: Benefits of the DoD Mentor-Protégé Program Are Not Conclusive*, July 2001. As of April 28, 2008:
http://www.gao.gov/new.items/d01767.pdf

————, *Best Practices: Improved Knowledge of DoD Service Contracts Could Reveal Significant Savings*, Washington, D.C.: GAO-03-661, June 2003. As of April 24, 2008:
http://www.gao.gov/new.items/d03661.pdf

————, *Contract Management: Impact of Strategy to Mitigate Effects of Contract Bundling on Small Business Is Uncertain*, May 2004. As of May 18, 2008:
http://www.gao.gov/new.items/d04454.pdf

General Dynamics Land Systems, "Supply Chain Management (SCM) Enterprise Requirements Planning (ERP) Oracle Implementation," letter, November 27, 2007. As of April 28, 2008:
http://procurement.gdls.com/pdf/112707_Oracle_Communication1.pdf

General Services Administration, *Federal Acquisition Regulation*, March 2008. As of April 24, 2008:
http://www.acqnet.gov/FAR/

Gerin, Roseanne, "Small Business Battlegrounds: Size Standards, Bundling, Women-Owned Firms Still Buzz Among Industry Issues," *Washington Technology*, September 26, 2005.

Global Computer Enterprises, Inc., "FPDS-NG Questions," 2006. As of May 25, 2008:
http://fpdsng.com/questions.html

Gottlieb, Daniel W., "Military Declares War on Spend," *Purchasing*, May 20, 2004. As of May 5, 2008:
http://www.purchasing.com/article/CA416336.html

Government Accountability Office, *Information on Awards Made by NIH and DoD in Fiscal Years 2001 Through 2004*, April 2006a. As of May 21, 2008:
http://www.gao.gov/new.items/d06565.pdf

————, *Best Practices: Stronger Practices Needed to Improve DOD Technology Transition Processes*, September 2006b. As of May 21, 2008:
http://www.gao.gov/new.items/d06883.pdf

————, *Small Business Innovation Research: Agencies Need to Strengthen Efforts to Improve the Completeness, Consistency, and Accuracy of Awards Data*, October 2006c. As of May 21, 2008:
http://www.gao.gov/new.items/d0738.pdf

————, *DoD Payments to Small Businesses: Implementation and Effective Utilization of Electronic Invoicing Could Further Reduce Late Payments*, May 2006d. As of May 22, 2008:
http://www.gao.gov/new.items/d06358.pdf

————, *Contract Management: Protégés Value DoD's Mentor-Protégé Program, but Annual Reporting to Congress Needs Improvement*, January 2007. As of April 28, 2008:
http://www.gao.gov/new.items/d07151.pdf

Hahn, Chan K., Kyoo H. Kim, and Jonj S. Kim, "Costs of Competition: Implications for Purchasing Strategy," *Journal of Purchasing and Materials Management*, Vol. 22, No. 3, Fall 1986, pp. 2–8.

Hardy, Michael, "OMB Looks to Grade Agencies on Bundling," *Federal Computer Week*, September 11, 2006.

————, "Code Shopping," *Washington Technology*, March 12, 2007. As of May 21, 2008:
http://www.washingtontechnology.com/print/22_04/30259-1.html

Hayward, Keith, "'I Have Seen the Future and It Works': The U.S. Defence Industry Transformation—Lessons for the UK Industrial Base," *Defence and Peace Economics*, Vol. 16, No. 2, April 2005, pp. 127–141.

Held, Bruce J., Thomas Edison, Shari Lawrence Pfleeger, Philip S. Anton, and John Clancy, *Evaluation and Recommendations for Improvement of the Department of Defense Small Business Innovation Research (SBIR) Program*, Santa Monica, Calif.: RAND Corporation, DB-490-OSD, 2006. As of May 6, 2008:
http://www.rand.org/pubs/documented_briefings/DB490/

Hewlett Packard, "HP Supplier Portal—Electronic Invoicing," 2008. As of June 5, 2008:
https://h20168.www2.hp.com/supplierextranet/invoicing.do

Hise, Phaedra, "The Remarkable Story of Boeing's 787," *Fortune Small Business,* July 9, 2007. As of June 5, 2008:
http://money.cnn.com/magazines/fsb/fsb_archive/2007/07/01/100123032/index.htm

Hofman, Mike, "When Defense Contractors Are Too Big to Be Small," *Inc.,* August 2006.

Hougui, Sadok Z., Aaron J. Shenhar, Dov Dvir, and Asher Tishler, "Defense Conversion in Small Companies: Risk, Activities, and Success Assessment," *Journal of Technology Transfer,* Vol. 27, No. 3, June 2002, pp. 245–261.

House Appropriations Committee, Department of Defense Appropriations Bill 2008, July 2007.

House Small Business Committee Democratic Staff, *Scorecard VII—Faulty Accounting by Administration Results in Missed Opportunities for Small Businesses,* July 26, 2006. As of May 12, 2008:
http://www.house.gov/smbiz/democrats/Reports/ScoreCardVIIFINAL.pdf

Howell, Elaine, Mentor-Protégé Division Chief, email correspondence with Nancy Y. Moore, RAND, June 16, 2008.

Innovation & Information Consultants, Inc., *Trends in Electronic Procurement and Electronic Commerce and Their Impact on Small Business,* Concord, Mass., June 2004.

Jennings, Jack, Adam Moody, Clyde Jackson, and Lou De Prospero, "A Review of the Department of Defense Pilot Mentor-Protégé Program," McLean, Va.: Logistics Management Institute, December 2000.

Jennings, John B., Janet Koch, and Clark Mercer, "Review of the DoD Mentor-Protégé Program," McLean, Va.: Logistics Management Institute, May 2006.

Johansson, Juliet E., Chandru Krishnamurthy, and Henry E. Schlissberg, "Solving the Solutions Problem," *The McKinsey Quarterly,* August 2003.

Johnson, Clay, III, "Implementing Strategic Sourcing," May 20, 2005. As of May 15, 2008:
http://www.whitehouse.gov/omb/procurement/comp_src/implementing_strategic_sourcing.pdf

Johnson, Larry, Policy and Performance Manager, Accounting Policy Division, DFAS Indianapolis, telephone interview, March 20, 2007.

King, David R., and John D. Driessnack, "Analysis of Competition in the Defense Industrial Base: An F-22 Case Study," *Contemporary Economic Policy,* Vol. 25, No. 1, January 2007, pp. 57–66.

Kopač, Erik, "Defense Industry Restructuring: Trends in European and U.S. Defense Companies," *Transition Studies Review,* Vol. 13, No. 2, July 2006, pp. 283–296.

Laseter, Timothy M., *Balanced Sourcing: Cooperation and Competition in Supplier Relationships,* San Francisco, Calif.: Jossey-Bass Publishers, 1998.

Lee, Deidre, *Update on the Transition to the Federal Procurement System—Next Generation,* August 1, 2005. As of June 23, 2008:
http://www.acq.osd.mil/dpap/policy/policyvault/2005-0969-DPAP.pdf

Liker, Jeffrey K., *The Toyota Way: 14 Management Principles from the World's Greatest Manufacturer,* New York: McGraw-Hill, 2004.

McCarthy, Ellen, "Contractors Cash Out but Try to Stay Humble," *Washington Post,* April 10, 2006. As of June 5, 2008:
http://www.washingtonpost.com/wp-dyn/content/article/2006/04/09/AR2006040900987.html?nav=rss_print/asection

Mead, Patricia, "Oversight and Management of the Government Purchase Card Program Reviewing Its Weaknesses and Identifying Solutions," May 1, 2002. As of June 23, 2008:
http://www.gsa.gov/Portal/gsa/ep/contentView.do?contentType=GSA_BASIC&contentId=13027&noc=T

Moore, Nancy Y., Clifford A. Grammich, and Robert Bickel, *Developing Tailored Supply Strategies,* Santa Monica, Calif.: RAND Corporation MG-572-AF, 2007. As of May 7, 2008:
http://www.rand.org/pubs/monographs/MG572/

Nelson, Dave, Rick Mayo, and Patricia E. Moody, *Powered by Honda: Developing Excellence in the Global Enterprise,* New York: John Wiley & Sons, Inc., 1998.

Nerenz, Timothy T., "Federal Government Procurement Policy Analysis: Has Extent and Effect of Contract Bundling on Small Business Been Overstated?" unpublished doctoral dissertation, Prescott Valley, Ariz.: Northcentral University, 2006.

————, "Government Contract Bundling: Myth and Mistaken Identity," *Defense Acquisition Review Journal,* Vol. 14, No. 3, December 2007, pp. 471–485. As of April 24, 2008:
http://www.dau.mil/pubs/arq/2008arq/ARJ46Web/ARJ46_Nerenz.pdf

Northrop Grumman Corporation, "My OASIS," 2008. As of June 5, 2008:
https://oasis.northgrum.com/myoasis.htm

Office of Management and Budget, Circular No. A-76 (Revised), May 29, 2003. As of April 30, 2008:
http://www.whitehouse.gov/OMB/circulars/a076/a76_rev2003.pdf

Office of the Under Secretary of Defense (Comptroller), *National Defense Budget Estimates for FY 2009,* Washington, D.C., March 2008. As of May 14, 2008:
http://www.defenselink.mil/comptroller/defbudget/fy2009/fy2009_greenbook.pdf

Office of the Under Secretary of Defense for Acquisition, Technology, and Logistics, "Electronic Subcontracting Reporting System (eSRS)," April 30, 2008. As of May 7, 2008:
http://www.acq.osd.mil/dpap/pdi/eb/electronic_subcontracting_reporting_system.html

Pae, Peter, "Acquirers Desire Secret Workers," *Los Angeles Times,* April 14, 2003.

Phillips, Bruce D., *Small Business Problems and Priorities,* Washington, D.C., and San Francisco, Calif.: National Federation of Independent Business Research Foundation and Wells Fargo, June 2004. As of June 5, 2008:
http://www.nfib.com/attach/6155

Public Law 85-536, Small Business Act, as amended through December 8, 2004. As of May 12, 2008:
http://www.sba.gov/regulations/sbaact/sbaact.html

Public Law 97-219, Small Business Innovation Development Act, 1982. As of June 4, 2008:
http://history.nih.gov/01docs/historical/documents/PL97-219.pdf

Public Law 102-564, Small Business Research and Development Enhancement Act, 1992. As of June 4, 2008:
http://history.nih.gov/01docs/historical/documents/PL102-564.pdf

Public Law 106-398, National Defense Authorization, 2000. As of June 5, 2008:
http://frwebgate.access.gpo.gov/cgi-bin/getdoc.cgi?dbname=106_cong_public_laws&docid=f:publ398.106.pdf

Public Law 106-554, Consolidated Appropriations Act, 2000. As of June 4, 2008:
http://history.nih.gov/01docs/historical/documents/PL106-554.pdf

Public Law 108-375, National Defense Authorization, 2004. As of June 23, 2008:
http://www.dod.mil/dodgc/olc/docs/PL108-375.pdf

Public Law 109-163, National Defense Authorization Act, 2006. As of June 4, 2008:
http://www.dod.mil/dodgc/olc/docs/PL109-163.pdf

Quaddus, Mohammed, and Glenn Hofmeyer, "An Investigation into the Factors Influencing the Adoption of B2B Trading Exchanges in Small Businesses," *European Journal of Information Systems,* Vol. 16, No. 3, July 2007, pp. 202 ff.

Reardon, Elaine, and Nancy Y. Moore, *The Department of Defense and Its Use of Small Businesses: An Economic and Industry Analysis,* Santa Monica, Calif.: RAND Corporation DB-478-OSD, 2005. As of May 12, 2008:
http://www.rand.org/pubs/documented_briefings/DB478/

Romney, Lee, "Strength in Numbers: Small Companies Team Up to Win Big Contracts," *Los Angeles Times,* May 27, 1998.

Small Business Administration, "Goaling Program Information," 2008a. As of May 3, 2008:
http://www.sba.gov/aboutsba/sbaprograms/goals/pi/index.html

————, "Fiscal Year 2007: Federal Small Business Prime Contracting Goals," April 10, 2008b. As of May 3, 2008:
http://www.sba.gov/idc/groups/public/documents/sba_program_office/goals_fy07_fed_small_goals.xls

———,"Guide to Size Standards," 2008c. As of May 3, 2008:
http://www.sba.gov/services/contractingopportunities/sizestandardstopics/indexguide/index.html

———, "About SDB Program: Eligibility Requirements," 2008d. As of April 28, 2008:
http://www.sba.gov/aboutsba/sbaprograms/sdb/program/sdb_aboutus_eligibility.html

Small Business Administration Office of Advocacy, *The Small Business Economy for Data Year 2006: A Report to the President,* December 2007. As of May 17, 2008:
http://www.sba.gov/advo/research/sb_econ2007.pdf

Smock, Doug, "P&G Boosts Leverage," *Purchasing,* November 4, 2004. As of April 23, 2008:
http://www.purchasing.com/article/CA479512.html

Statistical Information Analysis Division, "DoD Procurement Data," 2008. As of June 23, 2008:
http://siadapp.dmdc.osd.mil/procurement/Procurement.html

Teague, Paul, "Global Diversity Efforts Make Wise Investments," *Purchasing,* February 14, 2008. As of May 25, 2008:
http://www.purchasing.com/article/CA6530078.html

"Transcript of Debate Between Bush and Kerry, with Domestic Policy the Topic," *New York Times,* October 13, 2004. As of April 30, 2008:
http://www.nytimes.com/2004/10/13/politics/campaign/14DTEXT-FULL.html

Trent, Robert, J., "Applying TQM to SCM," *Supply Chain Management Review,* May/June 2001, pp. 70–78.

U.S. Census Bureau, "Statistics of U.S. Businesses," 2008. As of May 17, 2008:
http://www.census.gov/csd/susb/susb.htm

*U.S. Code Congressional and Administrative News,* 101st Congress, Second Session, 1990, v. 2., pp.104 Stat. 1,485–104 Stat. 1,855.

U.S. Senate Committee on Small Business and Entrepreneurship, "SBA in Hot Water for Failing to Monitor Contract Bundling's Effect on Small Business," July 18, 2005. As of April 30, 2008:
http://sbc.senate.gov/press/record_article.cfm?id=241733&&

Varmazis, Maria, "Supplier Diversity: Best Practices in Action," *Purchasing,* August 17, 2006. As of May 25, 2008: http://www.purchasing.com/article/CA6361158.html

Velázquez, Nydia M., letter to Laurie Duarte of the General Services Administration, April 1, 2003. As of April 30, 2008:
http://www.house.gov/smbiz/democrats/comments040103.pdf

Wal-Mart Stores, Inc., "Becoming a Supplier–Requirements," 2008. As of May 22, 2008:
http://walmartstores.com/Suppliers/248.aspx

Walter, Robert, Electronic Commerce Branch, DFAS Columbus, email to RAND, "RE: CDR," April 9, 2008.

WAWF Web-Based Training, home page, August 1, 2008. As of September 15, 2008:
http://wawftraining.com

Weigelt, Matthew, "Administration's Inaction Forces Action," *Federal Computer Week,* April 23, 2007.

Wessner, Charles W., ed., *The Small Business Innovation Research Program: Challenges and Opportunities,* Washington, D.C.: The National Academies Press, 1999.

———, *The Small Business Innovation Research Program: An Assessment of the Department of Defense Fast Track Initiative,* Washington, D.C.: The National Academies Press, 2000.

———, *An Assessment of the Small Business Innovation Research Program at the Department of Defense,* Washington, D.C.: The National Academies Press, 2007a (Prepublication Copy). As of February 14, 2008:
http://books.nap.edu/openbook.php?record_id=11963&page=R1

———, *SBIR and the Phase III Challenge of Commercialization: Report of a Symposium,* Washington, D.C.: The National Academies Press, 2007b.

"What's in a Name? New CPO Title Reflects Buyer's Strategic Role," *Purchasing,* June 1, 2000. As of May 7, 2008:
http://www.purchasing.com/article/CA138753.html

White, Bobby, and Vauhini Vara, "Cisco Changes Tack in Takeover Game," *Wall Street Journal,* April 17, 2008.

Womack, James P., and Daniel T. Jones, *Lean Thinking: Banish Waste and Create Wealth in Your Corporation,* New York: Simon & Schuster, 1996.

Womack, James P., Daniel T. Jones, and Daniel Roos, *The Machine That Changed the World,* New York: Harper-Perennial, 1991.

Zwahlen, Cynthia, "Tightened Rules Aim to Protest Small-business Federal Contracts," *Los Angeles Times,* March 28, 2007.